D0860251

PRIVACY AND THE COMMERCIAL USE OF PERSONAL INFORMATION

PRIVACY AND THE COMMERCIAL USE OF PERSONAL INFORMATION

by

Paul H. Rubin
Emory University and The Progress & Freedom Foundation

And

Thomas M. Lenard
The Progress & Freedom Foundation

KLUWER ACADEMIC PUBLISHERS
Boston / Dordrecht / London

Distributors for North, Central and South America:
Kluwer Academic Publishers
101 Philip Drive
Assinippi Park
Norwell, Massachusetts 02061 USA
Telephone (781) 871-6600
Fax (781) 871-6528
E-Mail <kluwer@wkap.com>

Distributors for all other countries:
Kluwer Academic Publishers Group
Distribution Centre
Post Office Box 322
3300 AH Dordrecht, THE NETHERLANDS
Telephone 31 78 6392 392
Fax 31 78 6546 474
E-Mail <services@wkap.nl>

Electronic Services <http://www.wkap.nl>

Library of Congress Cataloging-in-Publication Data

A C.I.P. Catalogue record for this book is available from the Library of Congress

Printed on acid-free paper.

Printed in the United States of America

Contents

Acknowledgements

This study is a product of The Progress & Freedom Foundation's project on *Regulating Personal Information: Balancing Benefits and Costs*. Needless to say, the project would not have been completed without the support of PFF. Many individuals commented on the manuscript at various stages. The authors would like to express their particular appreciation to Jennifer Barrett, Edwin Behrens, William Buzbee, Fred Cate, Fred deWolf, Jeffrey Eisenach, Charles Eldering, Tony Hadley, Dan Jaffe, Michael Hammock, Robert Litan, Mel Peterson, Solveig Singleton, Michael Turner and Victor Vornov for helpful comments and suggestions. Finally, we would like to thank Donna Anastasi, Brooke Emmerick, Katie Flint and Erik Heinecke for all the work they have done to produce a presentable manuscript and book.

Foreword

Hon. Orrin G. Hatch
United States Senate

Electronic commerce will be pivotal to the United States economy in the 21st Century. With the advent of electronic commerce, some consumers have become concerned about the disclosure, transfer, and sale of information which businesses have collected about them. These concerns purportedly are slowing the rate of expansion of electronic commerce, thereby putting at risk the future growth of the New Economy. To reduce this risk, a variety of schemes have been proposed under which the government would regulate online privacy. Congress currently is in the midst of a vigorous debate as to whether the government should regulate on-line privacy standards, and, if so, what form such regulation should take.

This succinct yet powerful book by Paul Rubin and Thomas Lenard goes to the heart of these issues. It explains that there is no evidence of actual consumer harm or market failure that could justify burdensome government regulation of online privacy. It describes the tremendous advantages consumers currently receive from the free flow of information collected on-line, advantages which could be eliminated if the government unnecessarily regulates and stops this flow of information. It argues that the free market provides businesses with compelling incentives to adopt their own measures — such as seal programs and novel technologies — to assuage consumer privacy concerns. This book presents compelling evidence to support these and many other points central to the continuing debate in the halls of Congress and elsewhere concerning online privacy.

Consumers must be confident that their privacy will be protected online if electronic commerce is to fulfill its full potential. Government regulation of personally identifiable information, however, raises many challenging issues that policymakers must get right if we are not to hinder the growth of electronic commerce. Because of its many insights, this book will be of great value to policymakers who must make tough choices about government regulation and to consumers who want to become better informed about online privacy issues.

This book is a must read for anyone who wants to make sure that the tough choices that government makes about online privacy are the right choices.

Executive Summary

The subject of this study is the commercial market for personal information and whether it should be subject to new regulation. Recent advances in information technologies have reduced the costs of gathering, storing, manipulating and transmitting information of all kinds. While the economic and social impacts of these advances have been overwhelmingly positive, they have also raised concerns on the part of individuals about what information is being collected, how it is being used and who has access to it. These concerns, in turn, have led to calls for new government regulation.

In order to decide whether regulation is desirable, and, if so, what form it should take, basic public policy questions need to be answered:

- Are there "failures" in the market for personal information?
- If market failures exist, how do they adversely affect consumers?
- Can such failures be remedied by government regulation?
- Would the benefits of government regulation exceed the costs?

The purpose of this study is to make a start toward answering these questions.

This study focuses on the market for personal information used for advertising and marketing purposes, which is the market affected by most of the regulatory and legislative proposals now under consideration.[1] While

[1] This study does not address specific categories of particularly sensitive information, such as health information, personal financial information or information about children. These types of information present separate issues and already are subject to regulatory programs specifically tailored for them. For example, the Children's Online Privacy Protection Act of 1998 (COPPA) regulates the collection, use and dissemination of personal identifying information obtained online from children under 13. Financial records have long been protected by the Fair Credit Reporting Act. More recently, the Gramm-Leach-Bliley Act of 1999 requires financial institutions to provide certain privacy safeguards to consumers for

much of the discussion concerns electronic information gathered over the Internet, the analysis applies to off-line information as well.

THE MARKET FOR PERSONAL INFORMATION

Data on individuals has been used by marketers and advertisers since long before the advent of the Internet. But, the Internet has increased the flow of personal information. This has produced large benefits, but it also has raised the level of individuals' concerns about privacy.

Targeted advertising on the Internet is accomplished by examining individuals' online activities, developing an understanding of their interests, and then matching and delivering relevant advertisements.[2] Advertisers compile individuals' Web-browsing activities and apply database technologies and statistical models that yield demographic and interest profiles. Advertisements relevant to consumers' profiles are then inserted into the Web pages they visit. Web site operators receive advertising revenues based on pages viewed and advertisements delivered.

CONSUMER BENEFITS

Both consumers and advertisers benefit from better targeting of advertising messages, which is made possible by the use of personal information. Consumers benefit from receiving information that is targeted to their interests, as well as from not receiving information that is not of interest to them. Apartment dwellers don't want information about aluminum siding, for example, and childless couples don't need to learn about infant formula specials. Similarly, marketers have an interest in not sending messages to consumers who aren't interested. Consumers are likely to avoid Web sites that routinely display information they find useless, and to ignore, delete or screen out messages from marketers who send the unwanted e-mails commonly described as "spam."

---

non-public personal information. In addition, the Health Insurance Portability and Accountability Act of 1996 (HIPAA) required the Department of Health and Human Services (HHS) to issue "Standards for Privacy of Individually Identifiable Health Information," which it did in December 2000. Similarly, this study does not cover government collection and use of information, which also presents a different set of issues. For an analysis of policies toward government records, see Alan Charles Raul, *Privacy and the Digital State*, The Progress & Freedom Foundation, 2001.

[2] Off-line advertising uses similar techniques.

Advertising revenues, which are made possible by the market for information, support a variety of valuable services that are provided to consumers at no charge. These services include free Internet access and e-mail, and pages from firms like Yahoo! customized to contain content of direct interest to the particular individual. Internet advertising firms, such as DoubleClick, provide customized advertising to smaller Web sites that use the advertising revenues to support themselves. Larger firms, such as AOL and Yahoo!, provide for themselves the same services that DoubleClick and its competitors provide for the smaller sites.

INFORMATION IS USED ANONYMOUSLY

Information used for marketing is generally sold in blocks of 1,000 consumers with a particular set of characteristics that makes them desirable to the marketer. Advertisers are not interested in the identity of individual consumers. An automobile company does not, for example, ask "What can I sell to Individual X?" It asks an advertising agency, such as DoubleClick or 24/7, to use its databases and statistical models to "place my ad on 1,000,000 pages viewed on computers of persons more likely than average to want a new car." Perhaps Individual X's computer turns out to be one of those selected, but no human makes this determination. Rather, it is made by computers connecting with each other.[3]

MULTIPLE USES OF INFORMATION

Once produced, information can be reused many times at low cost by different kinds of businesses. This "public good" characteristic is an important source of the productivity of information.

Advertisers, credit institutions and insurance companies all use the same commercial information discussed here. Indeed, the various information users cooperate in generating this information because they all find it valuable. Since the various uses of information subsidize each other, more information is collected and the cost to each of the users is reduced.

[3] A recent trend is for some companies to try to make their marketing better targeted and more personalized through Consumer Relationship Marketing ("CRM"), which generally tries to build on the relationship the company has with its existing customers. While companies engaged in CRM still market to a group of attributes rather than starting with an individual, they may end up marketing to an identifiable individual.

MARKET FAILURE AND HARM TO CONSUMERS

From an economic point of view, regulation of the market for information should be undertaken only if the market is not functioning properly. Market failure in this context would mean that consumers' preferences concerning the amount and use of their information are not being accurately transmitted and responded to in the marketplace. If the market is working well, there is no need for government to intervene. Regulation imposed on a market that is working well will not be helpful and, in fact, will introduce distortions.

After careful examination of the literature, we find no evidence of market failure or of harm to consumers from "too much" advertising and marketing information being produced. Given the widespread concerns about privacy and perceptions that personal information may be subject to misuse, it is noteworthy that there does not even appear to be much in the way of anecdotal evidence of harm to consumers from the legal use of information for marketing and advertising purposes. For example, in a year-end summary for 2000 dealing with privacy issues, CNET, a leading "New Economy" news source, indicated that there were no mishaps involving commercial use of personal information in 2000: "Despite the fears and concerns, there were no publicized horror stories that resulted from a privacy invasion." [4]

Implicit in the proposals to regulate the market for personal information is the view that there is a "market failure" resulting in "too much" information being produced, disseminated and used. As a general matter, however, markets work better with more information. As the cost of information goes down, market participants obtain more of it and, consequently, make better decisions. For example, if merchants can better estimate demand, they are less likely to purchase excess inventories, reducing costs and even lessening swings in overall economic activity. Similarly, geographic computer-based information can enable bricks-and-mortar merchants to put their new stores in the places that best serve consumers, and to stock the most useful merchandise for nearby consumers in those stores.

Electronic transmittal of information has led to a major reduction in the cost of information and therefore a major increase in the amount of information available to the economy. Since agents in the economy

[4] Patricia Jacobus, "Privacy heats up but doesn't boil over," CNET News, December 22, 2000, available online at http://news.cnet.com/news/0-1005-200-4238135.html?tag=st.cn.sr.ne.1, visited December 25, 2000.

generally benefit from better information, any policy that reduces the amount of such information below the efficient amount will have detrimental economic effects.

ASYMMETRIC INFORMATION

Asymmetric information is a form of market failure that occurs when one party to a transaction has more information than the other. Both credit markets and insurance markets are potentially subject to problems of this sort, because lenders and insurers may have less information than credit or insurance applicants about the applicants' risk characteristics.

An asymmetric-information market failure could exist on the Internet if consumers were unable to determine Web sites' privacy policies. But, mechanisms such as third-party "seal" programs are now available for firms to differentiate themselves and provide the necessary information.

In general, increased use of personal information solves, rather than exacerbates, asymmetric information problems. Moreover, the ability to sell for advertising or marketing purposes information initially collected for credit or insurance rating purposes increases the value of that information. Thus, the markets for advertising and marketing information generate increased information in markets that might truly be susceptible to asymmetric information – i.e., credit and insurance markets.

POSITIVE EXTERNALITIES

Market efficiency requires that buyers and sellers internalize the benefits and costs of transactions. If a transaction imposes costs on, or provides benefits to, third parties, then a market failure may exist. The classic example of a negative externality is environmental pollution. When the costs of pollution are not internalized by firms, they produce more of the polluting goods than is socially optimal.

A positive externality exists when an activity provides benefits to third parties. In this case, because decision makers do not consider the full benefits, they will undertake too little of the activity. Since information is a good that can be used repeatedly once it is produced, many activities involving information yield positive externalities.

There are also positive externalities associated with targeted advertising itself. The more unwanted messages consumers receive, the more likely they are to ignore all messages. This reduces the expected value of future messages from any sender. Thus, the ability of any individual advertiser to obtain the information necessary to better target advertising messages produces benefits to other advertisers – which is a positive externality.

Moreover, much of the information that sellers gather to target consumers is used for statistical modeling. Information of this sort needs to be collected from large numbers of consumers to be useful for statistical purposes. Provision of personal data by any individual, therefore, has benefits that go beyond the services provided to that individual – i.e., it enables other individuals to be better targeted.

MARKET REACTIONS TO PRIVACY CONCERNS

The absence of evidence of market failure should come as no surprise, because the market is visibly responding to privacy concerns in a variety of ways. First, firms that violate consumer expectations about privacy face a loss of "reputation" that translates into losses in the marketplace. Reputation effects are important, and the evidence shows that when a firm does something that is perceived as harming its reputation with consumers, the firm suffers a substantial loss in value.[5] We provide a number of examples of firms that have had to reverse course after adopting information policies that offended their customers. For example:

- Amazon discontinued its plans for "dynamic pricing" (what economists call price discrimination) when consumers learned about it and became irate.[6]
- America Online cancelled plans to sell subscribers' telephone numbers to telemarketers in response to angry reactions from subscribers.[7]
- Yahoo! eliminated the reverse telephone number search from its search site in response to consumer unhappiness.[8]

[5] For a summary of the literature, see Kari Jones and Paul H. Rubin, "Effects of Harmful Environmental Events on the Reputations of Firms," *Advances in Financial Economics*, 2001, edited by Mark Hirschey, Kose John and Anil K. Makhija, available online at http://papers.ssrn.com/paper.taf?ABSTRACT_ID=158849.

[6] David Streitfeld, "On the Web, Price Tags Blur," *Washington Post*, September 27, 2000. Amazon denied that it was engaged in dynamic pricing or price discrimination, but nonetheless suffered when consumers believed they were carrying out such policies.

[7] Jessica Litman, "Information Privacy/Information Property," 52 *Stanford Law Review*, 1283-1313, May, 2000, at 1305-6.

- Finally, there is the well-known story of DoubleClick's cancelled plan to link online and personally identifiable information through its acquisition of Abacus Direct.[9]

Thus, when businesses use information in ways that consumers do not like, consumers quickly learn about it, and the firms are forced to stop. Such reputational penalties may be among the strongest protections available to consumers. They send a powerful message to firms that they will incur losses if their information management policies are not to their customers' liking. Firms, therefore, have a strong incentive to avoid undertaking polices that run the risk of offending their customers. The Internet speeds up collection of information about consumers, but it also enables consumers to more easily obtain information about firms' activities on the Web.

Second, voluntary standards, defined and enforced by third parties or consortia of Web operators, are an important mechanism for providing information to consumers about Web sites' information policies. Many firms are adopting "seal" programs, whereby trusted third parties – TRUSTe, the Better Business Bureau, the Direct Marketing Association and others – certify the privacy policies of Web sites.[10]

There is evidence that voluntary standards in the U.S. actually work better than mandatory standards adopted by the European Commission.[11] About twice as many U.S. sites (62 percent) as European sites (32 percent) have posted privacy policies. Although all members of the EU now have data-privacy commissioners and agencies, these agencies seem unable to enforce privacy regulations. Thus, it appears that voluntary self-regulation provides more privacy protection than does mandatory government-imposed regulation.

Third, numerous new technologies, such as anonymous surfing and cookie control technologies, are available to consumers who are concerned about privacy. Consumers concerned about privacy are able to use these services, some free, to protect their information online.[12]

---

[8] Daniel J. Solove, "Privacy and Power: Computer Databases and Metaphors for Information Privacy," p. 27, available online through SSRN.Com, 56-57.

[9] Discussed at numerous places. See, for example, Diane Anderson and Keith Perine, "Marketing the DoubleClick Way," *The Standard*, March 13, 2000.

[10] Consumers can get information about these programs and links to other privacy sites at, for example, www.understandingprivacy.org.

[11] Ben Vickers, "Europe Lags Behind U.S. on Web Privacy: More American Firms Let Customers Guard Data, Study Finds," *The Wall Street Journal*, February 20, 2001.

[12] Some of these are discussed in Don Clark, "Privacy: You Have No Secrets," *The Wall Street Journal*, October 23, 2000 and Lorrie Faith Cranor, "Agents of Choice: Tools That Facilitate

Importantly, the World Wide Web Consortium (W3C), a consortium of 488 members (as of December 22, 2000), including the largest players on the Internet, such as Microsoft, America Online and Cisco,[13] is in the process of drafting a major private privacy protocol, the Platform for Privacy Preferences, P3P.[14] If P3P is successful, it will provide standardized information in machine-readable form about each Web site's privacy policy. Individuals will then be able to configure their own browsers to deal with the Web site.

REGULATION

Notwithstanding the absence of market failure and harm to consumers, there are proposals to regulate. The Federal Trade Commission bases its recommendations for greater regulation primarily on survey data, which, in general, are not a substitute for good public policy analysis.[15] Moreover, the opinion data that are available on privacy present an ambiguous picture concerning the public's desire for more regulation.

Regulating the market for information would raise its cost. As information about consumers became less available and more expensive, sellers would use less of it and rely more on sending messages to untargeted sets of consumers. Consumers would receive more unwanted messages and would, as a result, pay less attention to messages in general. This would increase the cost to producers of communicating with consumers and the cost to consumers of obtaining information. The loss in welfare would be significant.

FAIR INFORMATION PRACTICES

The FTC's proposed regulatory regime focuses on four fair information practices – Notice, Choice, Access, and Security. Some combination of these

---

Notice and Choice about Web Site Data Practices," available online from http://www.research.att.com/~lorrie/#publications.

[13] For the W3C homepage, see http://www.w3.org. For the list of members, see http://www.w3.org/Consortium/Member/List, visited December 22, 2000.

[14] http://www.w3.org/P3P/.

[15] Federal Trade Commission, *Privacy Online: Fair Information Practices in the Electronic Marketplace*, May 2000, available online at http://www.ftc.gov/os/2000/05/index.htm#22.

elements is incorporated in virtually all the legislative and regulatory proposals that are under consideration.

If regulation is to do any good, it must address harms caused by the use of information. In approaching this issue, it is useful to think about the information market as *two* markets. Information is acquired from individuals in one market, and processed and sold in a second market. If any harm results from the use of information, almost by definition this harm occurs in the second market, in which the processed information is used.

Many of the proposed fair information practices involve only the first market, however. This is particularly the case with Notice and Choice. Access also has little or nothing to do with the way information is used once it is acquired. Only Security deals with information at the point at which harm might occur.

NOTICE

Many Web sites, either through seal programs or otherwise, voluntarily provide notification about their information management practices. This is an appropriate market response to consumers' privacy concerns. A mandatory notice requirement, however, would be much less flexible and would likely impose substantial economic costs. Moreover, experience with existing mandatory notice requirements is not encouraging.[16]

A notice requirement would interfere with innovation in several ways. First, it would make it difficult, if not impossible, to take advantage of beneficial new uses of information. This is because a notice requirement is likely to require that consumers be notified whenever information is used for a purpose not initially intended. There are many examples of ways in which information is now being used that were not contemplated when the information was collected, and which might not be possible if a mandatory notice requirement had been in place.

Second, notice would likely impede the development of potential new uses of the Internet. The interaction of new technologies with privacy policies, including notice requirements, is problematic. For example, industry is responding to privacy concerns associated with new handheld devices, but would have a much more difficult time doing so if faced with a specific

[16] Consumers received notices from financial institutions required by the recently enacted Gramm-Leach-Blilely ("GLB") law and found them confusing and, in general, not useful. See John Schwartz, "Privacy Policy Notices Are Called Too Common and Too Confusing," *The New York Times*, May 7, 2001.

government mandate. There are a number of examples of regulations whose impact changes dramatically from one technology to another.

CHOICE

Choice also involves the market for acquisition of information, not the market for the sale and use of information, which is where potential harms may occur. As with notice, the only way choice can reduce misuse of information is by preventing information from being used at all. Recent studies suggest the cost of restricting data flows is high.[17]

A choice requirement could ironically deprive consumers of choices they now have available. A class of business models that provides consumers with "free" goods and services on the Internet, supported by revenues from advertising and the sale of information, may cease to exist. If this occurred, even consumers who would willingly exchange their information would not have the choice available. The services might continue if a sufficient number of consumers chose to continue providing information, but those who chose not to would free ride on the others.

Much of the choice debate revolves around the relative merits of opt-in and opt-out rules.[18] This decision should not, however, be made *a priori* by regulators. In some contexts, a Web site manager may choose opt-in because of the type of information at issue, the desires of the site's customer base or the value of the information to the site operator. In fact, some advertisers are willing to pay more for opt-in lists than for opt-out lists, on the theory that opt-in lists contain consumers who are more interested in receiving messages. Because different practices will be best for different situations, Web site managers should be able to configure their sites as they want, and consumers decide which sites to visit.

Moreover, it would be very difficult to design a regulatory regime that takes into account the complexities of the information flows in this market.

[17] Michael Turner estimates that opt-in requirements would reduce the data available to catalog apparel retailers at a cost of one billion dollars (in a $15 billion dollar market). See "The Impact of Data Restrictions on Consumer Distance Shopping," available at http://www.the-dma.org/isec/9.pdf. An Ernst and Young study conducted for the Financial Services Roundtable estimates that current information sharing by the 90 largest financial institutions provides benefits of $195 per household per year, for a total saving of $17 billion. See "Customer Benefits from Current Information Sharing by Financial Services Companies."

[18] These terms deal with the default for use of information. Under an opt-in rule, an individual must affirmatively agree to have his information used. Under opt-out, the gatherer of the information has the right to use it unless the individual requests that it not be used.

Theoretically, Web sites could be required to offer consumers choice with respect to each use of each piece of information. This would clearly involve a lot of choices and would place a significant burden on activities on the Internet.

An opt-in rule would dramatically reduce the amount of information available to the economy and impose substantial costs on consumers. The available evidence suggests that the great majority of consumers accept the default, probably because the transactions costs associated with making a decision are not trivial.[19] The efficient solution under these circumstances is to give the Web site the initial rights to the information. This is the transaction that consumers – who benefit from the Web site's use of their information – would make if transactions costs didn't get in the way. Thus, if there is any requirement at all, it should be opt-out.

ACCESS

Access allows a consumer to observe his information and perhaps correct errors. This is useful for information that is used personally, such as information used for credit checks and insurance applications. Its usefulness for marketing information is much less clear.

Implementing an access requirement would be extremely costly and would reduce the security of data on the Internet.[20] Indeed, there is a fundamental conflict between access and security. The FTC itself has been unable to define access in a meaningful way.[21]

SECURITY

Security does deal with misuse of information at the point where harm might occur. Thus, cost-effective security requirements, if they could

[19] In testimony before the FTC on the experience of one firm, a witness indicated that, when the default was opt-in, 85 percent of consumers chose not to provide their data. In contrast, 95 percent chose to provide their data when the default was opt-out. Testimony by Parry Ponemon, PriceWaterhouseCoopers, at the FTC hearing, "Wireless Web, Data Services and Beyond: Emerging Technologies and Consumer Issues," December 12, 2000, Vol. 2, p. 232.

[20] A recent study of online privacy regulation that focuses on access finds that costs "easily could be in the billions, if not tens of billions of dollars." See Robert W. Hahn, "An Assessment of the Costs of Proposed Online Privacy Legislation," May 7, 2001, available at http://www.actonline.org/press_room/releases/050801summary.asp.

[21] See Federal Trade Commission (2000a), *Final Report of the FTC Advisory Committee on Online Access and Security*, May 15, 2000, http://www.ftc.gov/acoas/papers/finalreport.htm.

be developed, might well be justified. There is, however, no reason to believe that firms now spend too little on security, since they have an interest in protecting their data and bear most of the cost of violations. Moreover, there is no indication that the FTC or any other government agency is better able than the private sector to provide for security. As with access, the FTC was not able to define security in a meaningful way, and indicated it would, if given the authority, determine a standard by adjudicating on a case-by-case basis.[22]

EFFECTS ON COMPETITION

Regulation that raises the costs of advertising and obtaining customer lists would have an adverse effect on new entrants, because advertising typically benefits new entrants and small firms more than it does large, established firms.[23] This is particularly true for Internet advertising, where established firms have lists of their own customers and visitors to their Web sites, but new firms must purchase such lists. As long as there is a market for customer lists and other such information, entrants can begin competing relatively easily. However, if regulation should reduce the size of this market and increase costs, competition from new entrants would be reduced.

Even notice requirements would disproportionately impact small businesses. The costs associated with notice are predominately fixed costs, and so are higher per unit of output for small than for large firms. Thus, any such regulations would serve at least in part as a barrier to entry against small firms, and as a source of protection for large established firms.[24]

FEDERAL VERSUS STATE REGULATION

The overall conclusion of our study is that regulation is not called for and would be costly for the economy. Given the nature of the Internet, regulation at the state level has the potential to produce additional costs and impede interstate commerce due to inconsistencies. Moreover, as in some other areas, states would have an incentive to adopt regulations that imposed costs

[22] See FTC Advisory Committee Report.
[23] John E. Calfee, *Fear of Persuasion: A New Perspective on Advertising and Regulation*, American Enterprise Institute, Washington, 1997.
[24] For the general point that firms may gain when costs of competitors increase, see Steven C. Salop and David T. Scheffman, "Raising Rivals' Costs," 73 *American Economic Review* (Papers and Proceedings) 267 1983.

on out-of-state producers. There is, therefore, a case for federal preemption of state regulatory authority.

CONCLUSION

The privacy debate presents some of the most complex policy-making challenges we have seen. It is disappointing, however, that so little careful analysis has been undertaken of the actual proposals and their likely consequences. This book attempts to start filling that gap.

We find that regulation imposed on a medium like the Internet that is changing so rapidly would have unpredictable consequences. The costs would take many forms. Regulation would limit the flow of information and make it more expensive. This could create market failures where none now exist. Perhaps the most serious cost would be a loss of innovation – new uses of information and of the Internet itself that would be frustrated by a new regulatory regime. All this would slow the progress of the IT revolution with potentially significant adverse effects on growth and productivity.

Chapter 1

Government, Markets and Privacy in The Digital Age

This book presents an evaluation of proposals to increase government regulation of the commercial use of personal information. Such proposals would require, for example, that companies provide specific notifications about their information management practices, or obtain the "opt-in" permission of consumers before collecting or using information about them. There is heated debate in Congress and elsewhere about the need for such regulation. Our goal is to inform that debate by applying widely accepted criteria to the analysis of government regulation: We ask, simply put, whether there are failures in the marketplace that result in consumer harm, and, if so, whether proposed new regulations would ameliorate these problems and yield consumer benefits greater than their costs.

It is important to recognize at the outset, however, that such proposals are part of a larger debate: the debate over privacy in the era of computers and the Internet. The proposals at issue in this book can only be understood fully in this larger context.

Technological progress inevitably produces social change, sometimes in small, subtle ways, other times at the deepest and most profound levels. From clocks to firearms to the splitting of the atom, new technologies have reshaped the ways people organize their daily lives, their economies and their political institutions. Whether or not such change is generally for the better is a matter perhaps more of taste than of resolvable debate. Whatever the balance, however, it is beyond argument that most important technological advances have complex effects which may be crosscutting or even contradictory.

Computers are no exception. As the most disruptive technological change since electricity, the digital revolution has, to use the economist's term, resulted in massive disequilibria, not just in markets for individual goods, but in institutions throughout society. One of those disequilibria is in the market for personal information, the subject of this book.

By making information more available, permitting diversity in both products and lifestyles, and in dozens of other ways, digital technologies expand choice and the ability of individuals to control their environment.[25] At the same time, by permitting the acquisition, retention and (potentially) the dissemination of vast amounts of data, computers hold the apparent potential to *reduce* individuals' control over their environment. This potential has been well understood at least since the publication of George Orwell's novel, *1984*. Such concerns have even motivated social protests, from the "Free Speech" movement at Berkeley in the 1960s, which led protesters to burn computer punch cards while shouting "do not fold, spindle or mutilate," to some of today's ardent advocates of personal privacy.

This tension is an important part of what the privacy debate ultimately is all about: Do computers and the Internet – or, more specifically, does the ability of both private and government institutions to gather and use increasing amounts of information about individuals – enhance or reduce individuals' control over their lives?

In thinking about this question, it is important to distinguish, first, between government and the private sector. Governments have the authority to mandate the collection of information and the power to use that information in ways that directly encroach upon our freedom. While government collection and use of information may have many positive effects – most would agree, for example, that mandatory disclosure of political contributions facilitates fair and open elections – concerns about government "spying on its citizens" are both deep-seated and, if history is any indication, not entirely without foundation.

To constrain such activities, the United States has enacted a complex and comprehensive web of Constitutional and statutory limits on the ability of governments to collect, maintain and release information about their citizens. Anchored in the Fourth and Fifth Amendments to the U.S. Constitution, these protections include landmark legislation, such as the 1974 Privacy Act, the 1980 Paperwork Reduction Act and the 1994 Drivers License Privacy Protection Act, as well as an extensive body of case law embodied in dozens of important court decisions. Despite these protections, many suggest that government continues to intrude on the privacy of its citizens, and there are

[25] See, for example, Esther Dyson, George Gilder, George A. Keyworth and Alvin Toffler, "A Magna Carta for the Knowledge Age," *Future Insight 1.2*, August 1994.

numerous proposals for legislation to further constrain its activities.[26] While this debate is not the subject of this book, it is clearly an important one.

The collection and use of information by private sector institutions – which is the subject of this book – poses a different set of issues. Unlike governments, private companies are not able to compel people to provide information. Consumers engage in market transactions of their own volition. No private company can compel you to give your e-mail address to the operator of a Web page. You do so only as part of a trade, which, at least on balance, you perceive as beneficial to you – or you would not do it.

This places important constraints on the power of private companies, constraints, which are not faced by government. We discuss these constraints at some length in this study. Simply put, however, market pressures force businesses to compete for the favor of customers, and subject them to consequences – lost business – if they do things that make customers unhappy. And that includes invading customers' privacy.

If this were all there were to it, this would be a short book indeed. Consumers would act only in their own best interests. Businesses, competing for customers, would do only things customers like. No regulation would ever be needed. But of course there is more to the story. Markets do not always work perfectly. Imperfect information, economic externalities and the existence of market power are examples of "market failures" that can lead to bad economic choices – that is, choices which, on balance, fail to maximize economic welfare. Moreover, the power of the marketplace to discipline private companies that fail to meet consumer needs is generally limited to withholding "repeat business"– in and of itself, it offers no solace or redress to consumers who may be harmed in the interim.[27]

For all of these reasons, governments can and do regulate information collection and use by private business. To date, such regulation generally has focused on areas where the potential for consumer harm is especially great, including health information, information about personal finances and information relating to children.[28] In each of these instances, it is the case

[26] For further discussion of these issues, see Alan Raul, *Privacy and the Digital State*, The Progress & Freedom Foundation, 2001.

[27] Although, obviously, firms will not knowingly engage in practices that offend a significant group of their customers.

[28] For example, the Children's Online Privacy Protection Act of 1998 (15 U.S.C. §§ 6501-6506) (COPPA) regulates the collection, use and dissemination of personal identifying information obtained online from children under 13. COPPA is enforced by the FTC. Financial records have long been protected by the Fair Credit Reporting Act. More recently, the Gramm-Leach-Bliley Act of 1999 (15 U.S.C. §§ 6801-6809) requires financial institutions to provide certain privacy safeguards to consumers for non-public personal information. And, the Health Insurance Portability and Accountability Act of 1996 (HIPAA) required the

that either consumers are in a relatively weak position to withhold information or the consequences of the misuse of the information potentially are especially large, or both. In these areas, in other words, the power of the marketplace to protect consumers and to discipline companies that cause harm is at its lowest.

We do not apply our analysis to these existing regulations, but do note two things about them: First, they are limited in scope and generally quite specific in their requirements. Second, despite this focus, they have sometimes proven more complex and difficult to implement than initially expected.[29]

The proposals at issue in this book are, by and large, far broader: Most of them would apply either to all commercial collection and use of information or, at a minimum, to all information collected or used "online." As discussed at length in this study, the degree of complexity involved in implementing such proposals and the attendant potential for large unintended effects are substantial.

Such disruptive effects could put at risk the undeniable benefits of electronic commerce. As Americans increasingly are aware, the Internet has transformed the process of retail purchasing.[30] It is now possible to get product information, compare prices and place orders, all with a few clicks of a mouse. The savings in terms of time and other resources are enormous. Consumers are able to obtain better information and transact at lower cost than ever before. They also enjoy the benefits of numerous free online services that would not be available without the support of Internet advertising revenues.

Department of Health and Human Services (HHS) to issue "Standards for Privacy of Individually Identifiable Health Information," which it did in December 2000. But these are not the only examples: existing laws prohibit the unauthorized release of school grade records by educational institutions (private and public alike); and, based on an incident in which opponents of Supreme Court nominee Robert Bork released copies of his video tape rental records, Congress prohibited unauthorized release of either video rental or cable television viewing records.

[29] When the Bush Administration took office in January 2001, there was substantial controversy over whether it should implement the medical-records privacy rules adopted by the outgoing Clinton Administration. Among the controversial items were provisions that might have blocked parents' access to their childrens' medical records or prevented relatives or friends from picking up prescriptions. Ultimately, the Bush Administration decided not to delay the new rules. See Laurie McGinley and Sarah Lueck, "Bush Won't Delay Medical-Privacy Rules Despite Objections by Hospitals, HMOs," *The Wall Street Journal*, April 12, 2001.

[30] Surveys done in 2000 indicate that between 75 and 135 million Americans were online. Jeffrey Eisenach, Thomas Lenard and Stephen McGonegal, *The Digital Economy Fact Book*, Second Edition, The Progress & Freedom Foundation, 2000.

Consumers are now spending about $26 billion per year over the Internet,[31] a figure expected to grow to about $270 billion by 2005, when the Internet will account for about eight percent of all retail sales.[32] The online-retail market is supported by online advertising, which is now between $8 billion and $9 billion per year and also growing rapidly.[33] These figures understate the importance of the Internet to retailing, because online and offline activities have become highly interdependent. On the one hand, traditional advertising leads to online purchases, with catalogues and direct mail advertising helping to drive consumers to retailers' Web sites.[34] On the other hand, consumers use the Internet to obtain product information and then undertake the actual purchases at traditional "bricks-and-mortar" outlets. This is especially the case for more expensive items, such as electronics and automobiles. Incorporating this effect brings the total amount of retail sales affected by e-commerce to an estimated $650 billion by 2005, or almost 19 percent of retail sales.[35]

Proposals to regulate the market for the very electronic information that has made e-commerce possible need to be evaluated carefully, especially because such markets evolve so quickly. As former White House e-commerce advisor Ira Magaziner put it, "[t]he Internet moves rapidly, mutating as technology changes. Governments inherently move slowly, bureaucratically – slower than is necessary for the Internet to flourish."[36]

The possibility for widespread effects on the economy is one important reason proposals to broadly regulate information use should be evaluated carefully. But it is not the only one. Another issue that is beginning to gain attention is the tension between our desire to protect our privacy, on the one hand, and the principles of freedom of speech and the First Amendment, on the other. As Professor Eugene Volokh has written, the right to privacy ultimately is a right to stop others from talking about you – a right we have seldom granted, especially for truthful speech. While our analysis does not extend to constitutional issues, we share the concerns expressed by some that

[31] For the year 2000. See http://www.census.gov/mrts/www/current.html.

[32] Yannis Bakos, "The Emerging Landscape for Retail E-Commerce," 15 *Journal of Economic Perspectives*, 69-80, Winter, 2001.

[33] Maryann Jones Thompson, "Scorecard for 2000," *The Standard*, March 19, 2001, at http://www.thestandard.com/article/display/0,1151,22825,00.html.

[34] This was a finding of research conducted by the Direct Marketing Association in cooperation with W.A. Dean and Associates. The DMA State of Catalogue/Interactive 2000, "The Convergence of Catalogue and E-Commerce," October 2000.

[35] Bakos, *op.cit.*

[36] Ira C. Magaziner, Creating a Framework for Global Electronic Commerce," *Future Insight*, The Progress & Freedom Foundation, July 1999.

privacy regulation must be seen in the full context of our civil liberties, and pay due deference to the First Amendment.[37]

Still another reason for careful evaluation of the new privacy proposals lies in the political environment in which they are being crafted and debated. There is no doubt that "privacy" is a hot political topic. Online privacy was the subject of eighteen major bills introduced in the 106th Congress (1999-2000), and privacy legislation was among the earliest bills introduced in the 107th Congress, in January 2001.[38]

The privacy issue is likely to remain important. *The Washington Post* indicates that privacy is one of the three most important Internet issues for 2001.[39] And, of nine legal experts asked by *The New York Times* to predict the most significant legal developments in Internet law and policy for 2001, five mentioned privacy.[40] Polling evidence, much of which is cited below, leaves no doubt that citizens are concerned about the impact of the digital revolution on their privacy.

There are reasons to believe, however, that such concerns are based in part on incomplete or even misleading information. Consider, for example, the debate over identity theft, the increasing incidence of which is sometimes cited as justification for the proposals at issue in this book. Identity theft is already a federal crime, and a crime in 42 states,[41] and the use of someone else's credit card is illegal in all 50 states. Moreover, despite the conventional wisdom, credit card fraud is declining.[42]

Somewhat surprisingly, the Internet does not appear to be a significant cause of these crimes. In a recent article, Betsy Broder, Assistant Director for Planning and Information at the FTC is quoted as saying: "The Internet is probably not as large a part of the problem [of identity theft] as people suspect." [43] Ms. Broder has also said, "None of the statistics show a greater

[37] For a complete discussion of these issues, see Eugene Volokh, "Freedom of Speech and Information Privacy: The Troubling Implications of a Right to Stop People from Talking About You," *Progress on Point 7.15*, October 2000 (previously published in the *Stanford Law Review*, Vol. 52, No.5, May 2000, pp.1049-1124).

[38] Patrick Ross, "Congress moves at light speed with Internet bills," CNET News.com, January 5, 2001.

[39] Christopher Stern, "New Congress Could Tackle Important Internet Issues," *Washington Post*, December 29, 2000. The other issues are taxation and copyright reform, but privacy is discussed first.

[40] Carl S. Kaplan, "Looking Forward," *New York Times* December 28, 2000.

[41] Identity theft state laws, available on the FTC Web site, http://www.consumer.gov/idtheft/statelaw.htm, visited March 15, 2001.

[42] Thomas A. Fogarty, "Credit Card Fraud Takes a Fall," *USA Today*, February 23, 2000.

[43] Quoted in Danielle Sessa, "The Best Way to ...Keep Safe," *The Wall Street Journal*, Nov. 27, 2000, R25.

vulnerability of consumers who are shopping online."[44] This is consistent with the findings of a study of 66 victims of identity theft, which found that only two of the 66 (about three percent) "had reason to believe that the thief had obtained their information via the Internet."[45] In another study, eight percent of the respondents who were victims of credit card fraud indicated that the thief "*might* have gotten the information because the consumer had provided it online."[46] The Inspector General of the Social Security Administration, James Huse has said, with respect to identity theft, "[t]his is not an Internet crime and never was."[47] A recent *New York Times* story indicates that "[d]espite the emphasis on security breaches on the Internet, most identity-theft cases start offline: a stolen wallet, pilfered garbage or a breached personnel file at a company's human relations office."[48]

Despite these facts, identify theft and other illegal activities are sometimes used as a rationale for supporting the new regulations at issue here. For example, consider the following two quotations:

> Anonymous tracking and profiling by DoubleClick and 24/7 [two leading Internet advertising agencies] can be very subtle. But sometimes privacy violations hit you in the face. We have all heard examples of sociopaths who stalk their victims online. We have seen the statistics on "identity theft," in which criminals suck enough personal data off the Net to impersonate other people.[49]

* * *

[44] Quoted in Susan Stellin, "Using Credit Cards Online Remains Safe Despite High-Profile Security Lapses," *New York Times* October 16, 2000.

[45] CALPIRG, "Nowhere to Turn: Victims Speak Out On Identity Theft," available on the CALPIRG Website, http://www.pirg.org/calpirg/consumer/privacy/idtheft2000/toppage1.htm visited January 12, 2001, p. 6.

[46] Susannah Fox and Oliver Lewis, *Fear of Online Crime*, Pew Internet and American Life Project, Pew Internet Tracking Report, April 2, 2001, available at http://www.pewinternet.org.

[47] Scott Bernard Nelson, "Identity Crisis," *The Boston Globe*, August 27, 2000. He does add: "But technology has created new ways of storing and selling personal information and it's likely to create more and more headaches in the future."

[48] Jennifer S. Lee, "Fighting Back When Someone Steals Your Name," *New York Times*, April 8, 2001. See also Robert O'Harrow Jr., "Identity Thieves Thrive in Information Age," *Washington Post,* May 31, 2110. This article indicates that, while the initial security breach typically is offline, identity thieves are using the services of commercial online data brokers.

[49] *Business Week*, "Online Privacy: It's Time for Rules in Wonderland," March 20, 2000, p. 83.

> Let's begin with a sense of the problem. Imagine that one day your bank or telephone company puts all of your transactions or phone records up on a Web site for the world to see. Imagine, more realistically, that the company without your permission simply sells your records to another company, for use in the latter's marketing efforts.[50]

Stalking, identity theft and unauthorized release of banking and telephone records "for the world to see" are offensive and, under existing law, quite illegal. They have nothing to do, however, with either online profiling or intercompany sales of marketing information, activities which would be proscribed or prohibited by the new regulations we examine here. These activities, and the impact the new rules would have on them, need to be analyzed in their own light, as we do below. It is crucial for informed decision-making by both citizens and policymakers that these distinctions be illuminated and clarified, not obfuscated and blurred.

From the profound issues of constitutional law and the First Amendment to the mundane imperatives of poll-driven politics, the "privacy" debate presents some of the most complex policy making challenges we have seen. It is disappointing that, at least until recently, so little careful analysis has been undertaken of the actual proposals and their likely consequences. This is especially the case with respect to the Federal Trade Commission's decision, in May 2000, to issue a report recommending new regulatory authority.[51] As discussed at length below, our analysis does not indicate that the new regulations the Commission requests would benefit consumers. This is, perhaps, not surprising, since the agency lacked a solid evidentiary or

[50] Peter Swire, "Markets, Self-Regulation, and Government Enforcement in the Protection of Personal Information," in Privacy and Self-Regulation in the Information Age, U. S. Department of Commerce, Washington, DC, 1997, http://www.ntia.doc.gov/reports/privacy/selfreg1.htm., at 1.

[51] Federal Trade Commission, *Privacy Online: Fair Information Practices in the Electronic Marketplace*, May 2000, available online at http://www.ftc.gov/os/2000/05/index.htm#22.

analytical basis for what it proposed. One Commissioner, Orson Swindle, went so far as to call the report "embarrassing."[52]

Our hope in this book is to begin filling the information gap that has, ironically, too often characterized the privacy debate to this point. We begin, in the following chapter, by looking at the market for commercial information, which is ground zero for analyzing any new regulatory regime.

[52] Hon. Orson Swindle, "Privacy in a Digital World: Industry Must Lead, or Government Will," *Progress on Point 8.4*, March 2001.

Chapter 2

The Market for Commercial Information

This chapter examines the functioning of the market for commercial information and the benefits it provides for consumers. The markets discussed here were important before the Internet, but the Internet has greatly increased their efficiency and scope.

We discuss three major interrelated sectors of the market for information: credit reporting, information aggregation and reselling, and Internet advertising. Although our focus is not on credit reporting *per se*, the credit reporting sector is important because it gathers information that is also used for marketing and advertising purposes. In fact, an important economic property of information is that, once produced, it can be used multiple times at a low marginal cost without any decrease in its value. This multiple-use property – that the same information is used for credit reporting and marketing purposes, for example – increases the social returns to information collection.

We describe how the three sectors gather, process and use personal information, and how this produces benefits for their customers and, ultimately, for consumers. The availability of personal information for accurate credit reporting is critical to the efficient allocation of credit. Information used for advertising and marketing provides consumers with a range of benefits.[53] Perhaps most importantly, the use of data for targeted marketing substantially reduces the cost of direct marketing, enabling firms to reach the right consumers and offer their products at lower prices than would be the case without access to personal information. As a byproduct, targeted advertising reduces the volume of unwanted messages ("spam"). In addition, revenue from advertising supports numerous free services, such as customized homepages and free Internet access and e-mail.

[53] Aggregators and resellers provide intermediate products, primarily to marketers and advertisers, and their benefits show up as benefits to those activities.

While personal data used by the credit reporting industry is, obviously, not anonymous, personal information used in the marketing and advertising sectors generally is used anonymously. The unit of commerce for advertising purposes is a block of consumers consisting of a large number of individuals with similar characteristics that are of interest to a specific marketer.

CREDIT REPORTING AGENCIES

Credit reporting and scoring has long been important to the efficient functioning of credit markets. The utility of credit reporting has increased in recent years as the Internet has increased access to information and reduced collection costs. This process is ongoing. The Internet increases the value of credit information by making it available to a greater number of users. This, in turn, increases the ability of credit scorers to gather additional information. One major benefit is that lenders are better able to tailor loan terms to individual borrowers.[54]

A credit reporting agency can broadly be defined as an aggregator of consumer credit information. These firms have diversified into related businesses, but credit reporting remains their core business. There are three major credit reporting agencies operating in the United States today: Equifax, Experian and Trans Union. These three agencies maintain very similar types of information on approximately 190 million individuals. As explained below, disclosure of this information is limited by provisions of the Fair Credit Reporting Act.

The credit reporting sector also includes more than 500 credit bureaus that either contract with or are owned by the principal reporting agencies.[55] These bureaus, which operate under various business models, serve as intermediaries and perform data gathering and distribution services.

Credit reporting agencies obtain data from credit-granting businesses, such as banks, credit card companies (Visa, MasterCard and retail store cards), other lenders and mortgage companies. These credit-granting institutions report their experience with the individuals to whom they have granted credit.

[54] Patrick Barta, "During Heightened Demand, Lenders Customize Home Loans," *The Wall Street Journal*, January 5, 2001, online edition.

[55] Number of individuals and number of credit bureaus from the Associated Credit Bureau website, http://www.acb-credit.com/, visited March 14, 2001.

The agencies add to this information publicly available information from public records. Historically, driver's registration and license databases have been an important source of information for current address and other personal information. These sources increasingly are restricted by the use of legally mandated opt-in procedures.[56] In fact, states have stopped selling the information, because of the administrative costs associated with opt-in. (We discuss the economics of opt-in requirements in greater detail in Chapter 6).

The data maintained by credit reporting agencies include name, address, Social Security number, telephone numbers, date of birth and employment information. This information is current and includes histories (previous addresses, employers, etc.). It also includes a detailed credit history– names of credit-granting institutions, types of credit, terms and payment history. The information available from public records might include tax liens, bankruptcies and other judgments that affect credit worthiness. The record also includes the names of entities that have made credit inquiries.

All this information is included in a credit report that, under the Fair Credit Reporting Act, can be sold only to entities that have a "legitimate business need." [57] A legitimate business need includes a response to a court order or a federal grand jury subpoena; and providing information in accordance with the written instruction of the consumer to a person who the credit agency has reason to believe: (a) intends to use the report in connection with the extension of credit or the review or collection of an account; (b) intends to use the report for insurance underwriting; (c) intends to use the report for determining eligibility for a government license or benefit where the government agency is required to consider the consumer's financial status; or (d) otherwise has a legitimate business need for the report in connection with a business transaction involving the consumer. The law prohibits the release of credit information to entities which do not meet one of these criteria.[58]

The credit agencies have continued to develop innovative service offerings as their business moves onto the Internet. While the traditional method of

[56] The Driver's Privacy Protection Act of 1994 (U.S.C. §§ 2721-2725) regulates the disclosure and resale of personal information contained in records maintained by state DMVs. An individual's information cannot be given out to credit reporting agencies, for example, without specific authorization (opt-in).

[57] A legitimate business need is defined in Section 604 of the Fair Credit Reporting Act (FCRA) available at http://www.acb-credit.com/.

[58] The Gramm-Leach-Bliley law, as interpreted by a recent court decision, appears to place additional limits on the dissemination of personal credit data. See Individual References Service Group, In. v. FTC, et al., Civil Action No. 00-1828, and Trans Union LLC v. FTC, et al., Civil Action No. 00-2087 (D.D.C. April 30, 2001). The implications of this decision for the commercial information market remain to be seen.

delivering printed credit reports is still available to fulfill the needs of clients that do not have access to computer networks, data are now being moved across the Internet, making use of secure network connections. This has facilitated real-time, on-demand, subscription-based credit-checking services. This permits credit grantors to automate real-time credit and risk-management decisions.

VALUE ADDED BY CREDIT REPORTING

The information provided by credit reporting agencies is used to allocate credit efficiently among potential borrowers. The agencies perform this function in several ways.

First, by providing information about the past history of borrowers, credit agencies create a "reputation" market. Borrowers who default on debt know that their borrowing information will become available to creditors in the future and their ability to borrow will be reduced. This increases their incentive to repay loans, which gives lenders increased confidence that their loans will be repaid.

Second, credit bureaus provide the information that lenders need to sort out borrowers and avoid problems of adverse selection.[59] If credit information were not available, some borrowers would borrow with no intention of repayment. Lenders would not have the information with which to identify those borrowers. As a result, little or no consumer lending would occur or, if it did occur, lenders would be forced to charge a large risk premium. It is estimated that the net benefits to consumers from accurate credit rating is a reduction in interest costs of 200 basis points (two percent) per year on average, which translates into $4,000 on a $200,000 mortgage, or $85 billion to $100 billion per year.[60]

Third, even borrowers with good intentions may be forced to default. But the probability of this occurring varies across borrowers. Lenders can gain by placing borrowers in the proper risk class and charging interest rates depending on risk. This means that good credit risks pay lower interest rates,

[59] See definition in glossary and discussion in Chapter 3. The basic theoretical argument is in Michael Rothschild and Joseph Stiglitz, "Equilibrium in Competitive Insurance Markets: An Essay on the Economics of Imperfect Information," 90 *Quarterly Journal of Economics*, 629-649, 1976.

[60] Walter F. Kitchenman, "US Credit Reporting: Perceived Benefits Outweigh Privacy Concerns," The Tower Group, Newton, Mass., 1998; Marty Abrams, "The Economic Benefits of Balanced Information Use," in *The Future of Financial Privacy*, Competitive Enterprise Institute, Washington, 2000.

and bad credit risks pay higher rates. This sorting leads to more lending and borrowing and produces net gains to consumers.

A similar set of arguments can be made about insurance markets. Increased personal information reduces problems of adverse selection in these markets, and therefore increases their scope and the associated consumer benefits.

INFORMATION AGGREGATION AND RESELLING

Numerous U.S. companies are engaged in the collection, processing and storage of data pertaining to individuals.[61] These firms obtain consumer information from credit bureaus, public records, telephone records, professional directories, surveys, customer lists and other data aggregators. The data are cleansed, using information on changes of name or address and death. Some firms distinguish between data collected under opt-in and opt-out procedures, with data provided under opt-in considered more valuable. These data are used to develop products, including telephone directories, household lists, and various specialty lists, such as lists of those who have recently moved. Resellers may rent these lists and perform various kinds of analyses for customers.

Information is gathered in a variety of ways. For example, Naviant Technologies captures purchasing and demographic information by providing services to companies whose customers register products on the Internet.[62] With its database of more than 17 million Internet-using households, Naviant helps clients develop traditional and Web-based direct marketing campaigns. Naviant also provides consulting and data processing services. In September 1999, Naviant signed an agreement with 24/7, one of the three major Internet advertising companies, which gave 24/7 access to Naviant's databases.

[61] A search of those firms classified in SIC 7374, "Companies engaged in marketing and business research services" yields approximately 50 companies.

[62] From Naviant's website, http://www.naviant.com/Products/elist/hthh.asp, visited January 1, 2000. Personal information includes data about: mail order buyers; purchasers of electronic/technology equipment; travel patterns; financial activities; sports and recreational activities; music and reading preferences; ownership of pets; and contribution/donation behavior. Demographic data include: zip code; telephone number; age, gender and family size; age of head of household; children's ages; income; length of residence; dwelling; Hispanic surname indicator; whether there is a home-based business; and the primary use of the purchase (business, personal or educational).

Information firms provide critical services to the insurance market. They are the source of the underwriting and claims information services needed to assess the insurability of individuals and businesses and to price individual and property insurance policies. The same firms that provide database-marketing services may also maintain databases of claims histories, and provide automated claims verification information services to the property/casualty and the life and health insurance markets.

Finally, there are specialized providers of information. A company called ChoicePoint, for example, maintains the largest database of physicians, chiropractors, dentists and orthodontists in the world.[63] This database is available online, where for a small fee it is possible to obtain information about a practitioner, including malpractice information and patient ratings. This is an example of value created by sharing information.

INTERNET ADVERTISING

Targeted advertising on the Internet (as elsewhere) is based on developing an understanding of consumers' interests, and then matching and delivering relevant advertisements. Internet advertisers develop consumer profiles by tracking an individual's online activities[64] and applying database technology and statistical models that yield demographic and interest profiles. These profiles are used to match advertisements with individuals. As the user visits the Internet, advertisements relevant to his or her profile are inserted into the pages the individual is requesting from the Web site. Advertising is a major source of revenue for Web sites such as search engines, directories and portals.

Internet advertising agencies deliver targeted advertisements across multiple Web sites. The three leading agencies are DoubleClick, Engage and 24/7 Media. DoubleClick, the largest online advertising agency, delivers advertising to approximately 1,300 of the most popular Web sites. Advertising agencies provide smaller Web sites with advertising revenue streams that they would not be able to obtain on their own.

Internet advertisers operate in the following manner. First, the advertiser interprets the immediate query the user makes. This involves using the Internet Protocol ("IP") address as a proxy for physical location, which

[63] http://www.choicepointinc.com.

[64] Actually, the information is about the browsing habits associated with a computer, which advertisers implicitly assume is used by an individual. Of course, multiple people in a single household may use the same computer.

provides information on the user's demographic profile. It also involves using the Web page that is being requested as an indicator of the individual's interests.

Second, the advertiser aggregates the queries that the user makes over time. To do this, advertisers make use of cookie technology that stores a unique identifier on the Web surfer's computer. The aggregation of queries provides the data used to statistically categorize Internet users into demographic or interest profiles. These profiles are used to match Internet users with relevant advertisements.

The flow of information from a viewer logging on to a Web site that participates in an Internet advertising agency's network might work as follows:[65]

1. The viewer visits the Web site, which checks the cookie on his computer. The Web site then displays an appropriate customized page.
2. The query is then referred to the agency, whose computer checks for its cookie and matches its unique identifier to the individual's profile on the agency's profile server. The profile is based on prior Web surfing that this particular individual has undertaken. Through the use of its cookie, the agency has been able to track across all sites that participate in its network.
3. The profile, which describes the person as a type, is forwarded to a correlation server where the viewer is matched with an advertisement. The agency's adserver delivers an ad into the original page delivered by the Web site.
4. After the advertisement has been delivered, the adserver can generate reports that provide the basis on which the advertiser – the firm whose product is being advertised – can be billed for the completed campaign. The agency can then deliver to the Web site its share of the revenues.

As discussed in more detail below, advertisements are typically delivered and sold in blocks of 1,000 viewers, maintaining the anonymity of individual

[65] This is a generic example of online advertising and does not necessarily track the exact methods used by any one agency.

consumers. Advertisers hire advertising agencies, such as DoubleClick, to find blocks of consumers who are more likely than average to purchase their products. This process does not involve examining the characteristics of any individual consumer to determine which products that consumer would buy.

The pricing policies and marketing programs of online advertisers show that more information about consumers is worth more money to advertisers. This is the case because better information can be used to more precisely target advertisements to consumers who have a higher probability of purchasing the good or service the advertiser is trying to sell. The higher value to the advertiser also reflects a higher value to the consumer.

DOUBLECLICK

The operations of DoubleClick provide additional detail on the way the Internet advertising industry works. DoubleClick's basic business is putting "banner" ads on Web sites. The company segments viewers of its ads in several ways.

For its basic programs, DoubleClick provides several "networks," which are lists of Web sites segmented according to subject matter, such as automobiles, business and finance, commerce, entertainment, technology, travel, women and health. These categories are based on information about the Web sites, not the viewers. The automobile network, for example, includes such Web sites as Kelley Blue Book, Autobytel.com, Popular Mechanics, and Automobile Magazine, and is aimed at viewers who are interested in cars. Similarly, the other specialized networks aim at viewers who are likely to be interested in particular classes of products.

Once an advertisement is placed on a network it can be further targeted to particular types of consumers, for an additional fee. The cost per thousand (CPM) viewers of the advertisement increases as the viewers are filtered based on additional characteristics. A typical CPM is $30 to $50, depending on the network and other characteristics. This information is, thus, worth three to five cents per use.[66] (That is also the approximate cost of a targeted e-mail sent for promotional purposes; such e-mails cost between one and five cents.[67]) The CPM increases by about one dollar for each additional filter that is applied. This implies that an additional piece of information about a consumer is worth one-tenth of a cent for each use by an advertiser.

[66] $30 per thousand viewers/1000 = three cents per individual viewer.

[67] Thomas E. Weber, "Why Companies Are So Eager To Get Your E-Mail Address," *The Wall Street Journal*, February 12, 2001.

DoubleClick uses four types of filters or targeting mechanisms: behavioral targeting; tech targeting; user targeting; and DoubleClick Intelligent Targeting.[68] These mechanisms can be applied to all of DoubleClick's sites, to its "Select" sites, or to any of the networks mentioned above. The Select sites are "a collection of the Web's premier sites" represented exclusively by DoubleClick. Thus, for example, a firm could advertise automobiles on the auto network for $30 per CPM; on the auto Select sites for $50 per CPM; or for targeted viewers on either of these sets of sites, for an additional charge of about one dollar per CPM for each filter applied.

DoubleClick also obtains information from viewers during the Internet viewing process. For example, it can target ads based on particular terms used in Web searches (e.g., stockbroker ads for viewers searching terms such as "stock" or "investments"). It can also monitor click-through behavior in order to determine in real time, as ads are being run, criteria for successful advertising. It can relate click-through rates to many of the specific filtering categories mentioned above, so that ads can be better targeted to responsive viewers. In addition to banner ads, DoubleClick also runs e-mail ad campaigns for clients. Again, these campaigns can be directed to specific types of viewers who are more likely to respond to a particular campaign. DoubleClick also has an opt-out privacy policy with respect to its cookies.[69]

OTHER ADVERTISERS – 24/7 AND ENGAGE

The other major Internet advertising agencies work the same way. In all cases, increased information about consumers translates into increased value. For example, 24/7 charges $40 per CPM or more for single site advertising, and as little as ten dollars per CPM for run-of-network advertising. It also

[68] Behavioral targeting allows advertisers to select audiences based on patterns of Web usage. The basic subcategories are psychographic targeting, where users are those most likely to respond; frequency control, which limits the number of exposures per individual to the ad; and hour-and-day control, which controls the time of day when the ad is seen. User targeting allows the advertiser to select the type of viewer, again in groups of 1,000. Subcategories include geographic (county, state or zip code); domain (.com, .edu, or .gov); company name; domain registration; industry type, based on the SIC code of the industry determined from the domain registration; and business size. Tech targeting allows advertisers to select viewers by user hardware, software, and Internet access provider. For example, engineers are more likely to use UNIX systems; graphic designers are more likely to use Macintosh systems. DoubleClick Intelligent Targeting aims at targeting the most responsive audience, based on interest and gender characteristics. Additional categories are being developed.

[69] We discuss market responses to privacy concerns in Chapter 4, below, and the economics of opt-in/opt-out in Chapter 6.

runs e-mail campaigns. In these campaigns, the CPM again varies with the degree of targeting. Interestingly, 24/7 advertises that its e-mail lists are opt-in. Advertisers view opt-in lists as being more valuable than opt-out lists, and are willing to pay more for access to such lists.

Engage also charges higher rates for more targeted ads – that is, ads using more consumer information. Engage has developed various profiles of viewers by interest.[70] Engage charges between $20 and $30 per CPM for campaigns based on profiling, and as little as five to six dollars per CPM for non-optimized run-of-network advertising.

PORTAL SITES

Portal sites, such as Yahoo!, America Online, Netscape, Microsoft Network, and CNN offer advertising services similar to those offered by the Internet advertising agencies discussed above. But, they offer a range of additional services as well. Their ability to provide these services is heavily dependent on advertising revenues.

Yahoo! provides a good example of how these sites work. Consumers provide Yahoo! with personal information, which can be exchanged with and used by Yahoo!'s business partners. In return, Yahoo! provides access to free information services offered through a personalized home page, MyYahoo!. Service offerings include free e-mail, calendar, contact database, and customized content, such as weather, stock quotes, news, sports scores and entertainment.

In order to gain access to the personalized services, the user must submit personal information, including name, country of residence, zip code, date of birth, gender, occupation, and industry of employment. The user can also provide optional information about areas of interest. These personalized data might also be augmented with credit agency data, or used in conjunction with profiles from Internet advertising agencies in order to better target advertising to Yahoo! users.

Yahoo! finances the provision of services to its users by selling advertising, which accounts for 90 percent of its revenue.[71] Every major Internet advertiser places ads on Yahoo!. Without the personalized data that

[70] These interests include: affluent interests; business professionals; computer enthusiasts; computer software enthusiasts; entertainment fans; food and wine connoisseurs; gamers; investors; music aficionados; online shoppers; outdoor and athletic enthusiasts; parents and families; sports fans; travelers; and young adults.

[71] http://yahoo.marketguide.com/mgi/MG.asp?nss=yahoo&rt=qbussegm&rn=A0DCA.

Yahoo! obtains from its subscribers and elsewhere, this business model would not work and the personalized services that Yahoo! and similar sites provide to consumers would not be available.

As with the Internet advertising agencies, the CPM that Yahoo! charges for the ads it sells varies as the ads become more targeted. For example, for run-of-category ads with demographic targeting, prices range from $31 to $156 per CPM. For untargeted run-of-category ads, the range is from $24 to $112 per CPM.

INFORMATION IS USED ANONYMOUSLY

As previously mentioned, information used for marketing and advertising purposes generally is used anonymously. Advertisers have no interest in the identity of individual consumers. An automobile company, for example, does not ask "What can I sell to Individual X?" Rather, it asks an advertising agency, such as DoubleClick or 24/7, to use its databases and statistical models to "put my ad on a million pages viewed on computers of persons more likely than average to want a new car." [72] Perhaps Individual X's computer turns out to be one of those selected, but no human makes this determination. Rather, it is made by computers connecting with each other. Moreover, the unit of commerce in the online advertising market is 1,000 persons, not any individual.[73]

This point is borne out by a discussion from an industry magazine about the process of profiling:[74]

> mind of how it works: A T-shirted techie huddles in the glow of his monitor, vigilantly tracking your every move on the

[72] A recent trend is for some companies to try to make their marketing better targeted and more personalized through Consumer Relationship Marketing ("CRM"), which generally tries to build on the relationship the company has with its existing customers. For example, an automobile company may search its customer database to identify those individuals whose leases are expiring and who have experienced service problems. These customers may then be contacted individually, to try to get them to renew their leases. Thus, while companies that are engaged in CRM still market to a group of attributes rather than starting with an individual, they may very well end up with an identifiable individual.

[73] "All rates are expressed in cost per thousand (CPM) ad banner impressions." From DoubleClick's Rate Card, http://www.doubleclick.net:80/us/advertisers/media/network info/rate-card.asp?asp_object_1=& , visited February 22, 2001. Other advertising agencies such as 24/7 have similar statements.

[74] Margaret Barnett, "The Profilers: Invisible Friends," *The Industry Standard*, March 13, 2000, p. 220.

> Net. Suddenly, he strikes. "He's at ESPN.com!" he shouts. "Send him the Chevy pickup ad!"
>
> The truth about targeting is much less exciting. But it's far more compelling. Someone is not watching you, something is: the all-seeing, all-knowing eyes of a bunch of Windows NT servers, humming quietly to themselves in a cool, dark room.

Personal information is held in "a bunch of Windows NT servers." It is not "known" in any useful sense by a human. A similar point is made in a paper by Daniel Solove:[75]

> Since marketers are interested in aggregate data, they do not care about snooping into particular people's lives. Much personal information is amassed and processed by computers; we are not being watched by other humans, but by machines, which gather information, compute profiles, and generate lists for mailing, e-mailing, or calling. This impersonality makes surveillance less invasive.
>
> While having one's actions monitored by computers does not involve immediate perception by a human consciousness, it still exposes people to the possibility of future review and disclosure. In the context of databases, however, this possibility is remote. Even when such data is used for marketing, marketers merely want to make a profit, not uproot a life or soil a reputation.

Thus, consumers' fears of loss of privacy appear to be based, at least to some extent, on a lack of understanding about the way in which their personal data are collected and used for advertising purposes. Their information is "known" only in the sense of being available on some computer.

[75] Daniel J. Solove, "Privacy and Power: Computer Databases and Metaphors for Information Privacy," p. 27, available online through SSRN. This paper is not published, but has been the subject of a *New York Times* story: Carl S. Kaplan, "Kafkaesque? Big Brother? Finding the Right Literary Metaphor for New Privacy," February 2, 2001. Solove is still in favor of regulation, however, and he perceives other dangers from online information, although the harm is "difficult to describe" and "difficult to quantify" (p. 40).

CONSUMER BENEFITS FROM ADVERTISING

As the discussion above indicates, consumers receive numerous benefits from Internet advertising, which, in turn, is made possible by the availability of personal information.

First, advertising – especially targeted ads made possible by the availability of personal data – provides valuable information to consumers.[76] If a consumer makes a purchase as a result of receiving an advertisement or solicitation from a Web site, then that information has been valuable to the consumer. The modern theory of advertising[77] indicates that most or all advertising provides valuable information, and if advertising leads to sales then at least some subset of consumers is benefiting from the advertising. Even if some information does not lead directly to a purchase, the information may still be valuable to the consumer. For example, it may better enable the consumer to compare prices among products, or to determine what products are available.

As advertising becomes more targeted, recipients avoid the nuisance of receiving unwanted ads that are of no interest to them. In the marketing literature, these are called "consumer-borne marketing costs".[78] Thus, both consumers and advertisers have incentives to more accurately target messages. Consumers gain because they receive fewer unwanted messages, and advertisers gain because consumers pay more attention to their messages.[79]

Second, advertising revenues support many valuable services provided to consumers at no charge. These services include customized pages from firms like Yahoo!, which contain content of direct interest to the particular individual, free Internet access as well as free e-mail services from many providers.

[76] For an important discussion of the value of information from an economic perspective, see Carl Shapiro and Hal Varian, *Information Rules: A Strategic Guide to the Network Economy*, Cambridge, MA: Harvard Business School Press, 1999.

[77] Summarized in Paul H. Rubin "Economics and the Regulation of Deception," *Cato Journal*, 1991,667-690, 1991 and John E. Calfee, *Fear of Persuasion: A New Perspective on Advertising and Regulation*, American Enterprise Institute, Washington, 1997.

[78] Ross D. Petty, "Marketing Without Consent: Consumer Choice and Costs, Privacy and Public Policy," 19 *Journal of Public Policy and Marketing*, Spring, 2000, 42-53, at 42.

[79] Ross D. Petty, "Marketing Without Consent: Consumer Choice and Costs, Privacy and Public Policy."

All of the free content available on the Internet is supported by advertising revenues. The major Internet advertising agencies work with operators of lower-volume Web sites to provide customized advertising, the revenues from which support the Web sites. Larger firms, such as AOL and Yahoo!, provide themselves the same services that DoubleClick and its competitors provide for smaller sites. DoubleClick and other Web advertising agencies enable the smaller Web sites to compete with the larger firms that have their own customer lists.

MULTIPLE USES OF INFORMATION

One of the most important economic characteristics of information is that the same information can be used in different ways by different parties. Once the initial costs of gathering and "producing" the information are incurred, additional uses can be undertaken at a relatively low marginal cost. This is an important source of value. Indeed, as illustrated below, advertisers, credit institutions and insurance companies all use the same commercial information and cooperate in its generation.

While different types of firms use the same information, the nature of the use is quite different. Credit agencies and insurance companies are interested in data about particular individuals. When an individual applies to buy insurance or borrow money, the insurer or lender wants to obtain data about that individual relevant to the insurance or credit decision. Advertisers, on the other hand, are not interested in data about individuals. As discussed above, they purchase data about blocks of individuals who share a set of attributes.

It is sometimes argued that information should be used only for the purpose for which it was initially collected. This is part of the European Union Directive on the Protection of Personal Data, and is advocated by some as appropriate policy for the U.S. as well. Such a restriction would be particularly costly because of the "multiple-use" characteristic of information, and would preclude low-cost, high-value uses of information that could otherwise occur. This issue is discussed in greater detail in Chapter 6.

Figure 1 illustrates how multiple parties – credit institutions, insurance companies, and direct marketers – all use the same information. Begin with an individual applying for credit, path A in Figure 1. Link A1 indicates that the individual grants the lending agency authority to request a credit report (link A2). The Credit Bureau obtains information from the Credit Reporting

Agency (link A3). The Credit Reporting Agency can obtain information from Information Aggregators (link B3) who themselves obtain information from Public Records (link B1). The Credit Reporting Agency combines this information and reports back to the Credit Granting Institution (link A5) and a final decision is made to grant or deny credit (link A6).

Path C represents information flows through Industry Specific Associations, such as Catalog Retailers (an example of a Direct Marketing Entity). Catalog Retailers periodically pool their data (link C1) and the Industry Specific Association then compiles this information and maintains industry-specific aggregate profiles. Retailers then purchase lists developed to their target market specifications (link C2). To augment and improve the quality of information purchased from the association, retailers may turn to either credit reporting agencies or data aggregators, such as Acxiom (link C3).

Path D represents an individual applying for insurance, D1. The Insurance Company will then request information from Insurance Industry Specific Associations, with whom it also shares information. This is shown in links E1 and E2. The Insurance Agency may also request information from the Credit Reporting Agency to enhance its risk management decision making. The information flows here follow the same path as discussed above.

Figure 1. Multilateral Information Flows

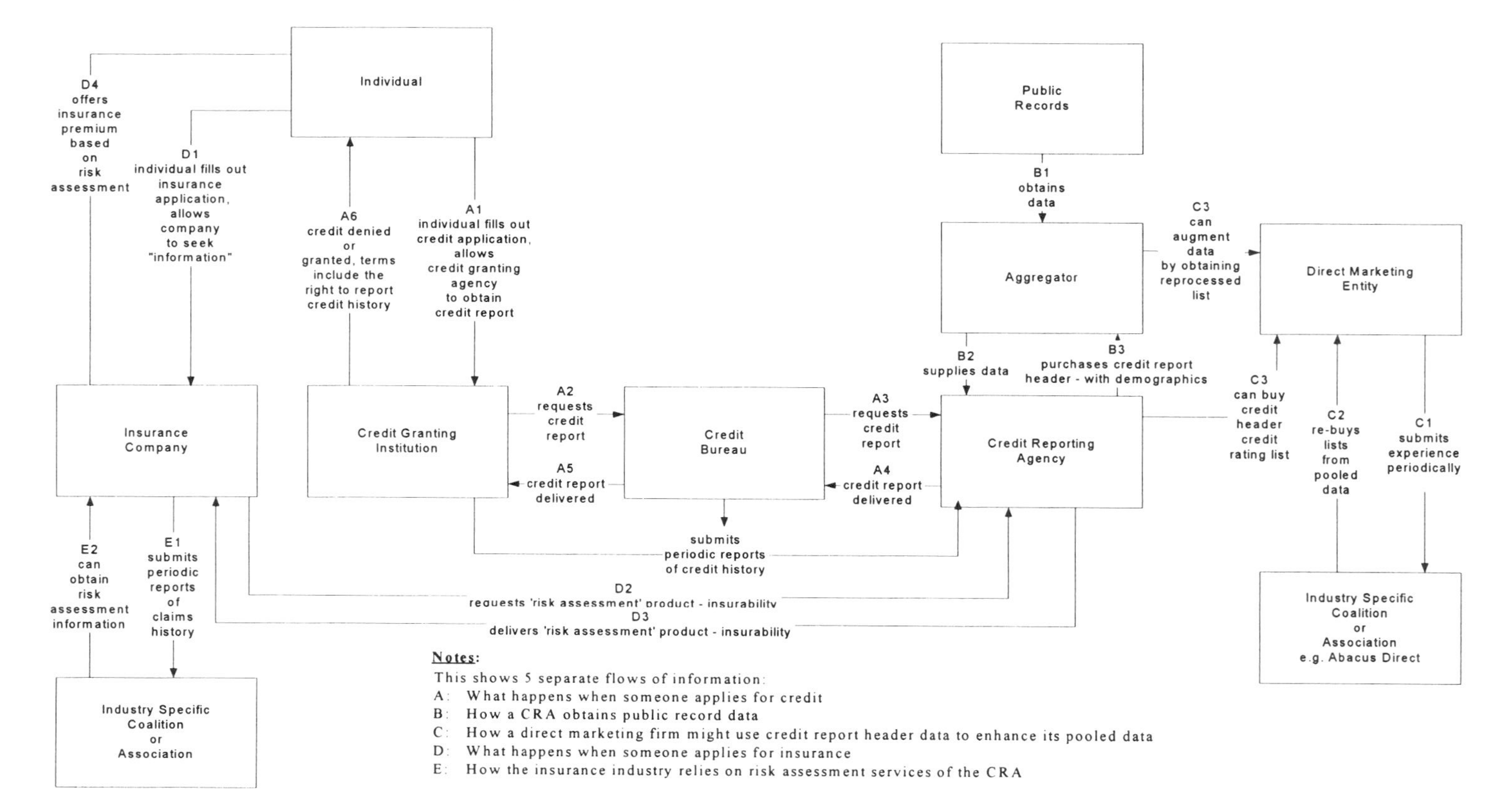

SUMMARY

Targeted advertising, which is made possible by the availability of personal information, provides many benefits. It provides information about products and services that are likely to be of interest to consumers. It pays for "free" services, such as free Internet access and e-mail and customized homepages. Finally, it reduces "spam" and other unwanted advertising messages.

Information about consumers comes from a variety of sources, and the same information is used for a number of different purposes. Information may be used to determine credit risks, price insurance, and for marketing purposes. Since information can be reused numerous times at low marginal cost, the social returns to collecting it are high. Moreover, since the various uses of information subsidize each other, more information is collected. This means that the cost of each of the uses is lower than otherwise.

In advertising, information is used anonymously. The unit of commerce is a block of 1,000 viewers who share a set of attributes that makes them desirable to the marketer. Advertising information is used and manipulated by computers, not human beings. Metaphors about Web sites "tracking" individuals are misleading.

Chapter 3

Market Failure and Consumer Harm

Markets can "fail" for a variety of reasons and the result is that either "too much" or "too little" of a good is produced. Implicit in the proposals to regulate the market for personal information is the view that there is a failure in the market for personal information resulting in "too much" information being produced, disseminated and used. Economic welfare would be improved if market failures were corrected and the "right" amount of goods produced.

In the absence of evidence of market failure, regulation is not justified– at least from an economic perspective. Regulating a market that is working properly – i.e., that is not failing – will produce distortions that impose costs on consumers. In the case of the market for information, imposing regulation on a properly functioning market would reduce the amount of information below the optimal level. If there are positive externalities, regulation would reduce information flows below levels that are already sub-optimal.

If a market is failing, there must be some evidence that the operation of the market is imposing some costs or harm on consumers. The opposite is not true, however. The existence of some harm does not necessarily mean that there is a market failure justifying an increase in government intervention. If there is evidence of harm, then the costs and benefits of proposals to reduce that harm need to be evaluated on a case-by-case basis to determine whether their benefits exceed their costs.

This chapter examines the potential for market failure and harm in the commercial information market. To the extent that there is a market failure, we argue that it results in an underproduction – not an overproduction – of information. This is because there are positive externalities associated with information.

We find no evidence of harm to consumers. While there obviously are examples of consumers being harmed by breaches of security and by information falling into the wrong hands, these incidents are not related to advertising, the collection of information by Web sites, or the commercial use of information.

There are three major forms of market failure – asymmetric information, externalities and monopoly power – which we discuss in turn. We then discuss the evidence concerning consumer harm.

ASYMMETRIC INFORMATION

The problem of asymmetric information occurs when one party to a transaction has more information than the other. In a classic analysis, George Akerlof used the example of the market for used cars.[80] The situation he posited was that sellers would know the quality of used cars, but buyers would not. Assume that a certain car – for example, a 1999 Chevrolet Malibu – is worth $15,000 in average condition, $18,000 in excellent condition and $12,000 in poor condition. If buyers do not have reliable quality information, they will pay no more than $15,000, which is the average value of such cars. A seller of a car in excellent condition knows, however, that it is worth $18,000 and, therefore, will not be willing to sell if he can get only $15,000. This implies that only average and poor quality cars will be offered on the market. But buyers, knowing that there are no excellent cars, will pay no more than $13,500 – the average price for cars that are in either average or poor condition. This, in turn, means that owners of average cars, worth $15,000, will not offer them, since they can get only $13,500. In the end, the only equilibrium is for cars that are in poor condition – "lemons" – which will sell for $12,000.

This is a market failure. The market for excellent and average cars, which should exist, doesn't, because of asymmetric information – sellers know the condition of the cars and buyers do not. The market fails because, despite the possibility of value-increasing trades between buyers and sellers of these cars, such transactions do not occur.

This example can be generalized, and there are many situations in which asymmetric information can theoretically lead to market failure. For example, both credit markets and insurance markets are potentially subject to market failures of this sort, because lenders and insurers have less information about applicants' risk characteristics than do the applicants themselves. In both cases, additional personal information about the applicants can help alleviate the problem.

[80]George A. Akerlof , "The Market for 'Lemons': Quality Uncertainty and the Market Mechanism," 84 *Quarterly Journal of Economics*, 1970, 488-500.

An asymmetric-information market failure could exist on the Internet if consumers were unable to determine Web sites' privacy policies. Under this circumstance, a "lemons" market, with no site providing privacy protections, could in theory come into being. The market would not provide useful information about privacy policies, which, in turn, would make some consumers reluctant to undertake transactions online.

The Federal Trade Commission believes that a market failure of this sort exists in the Internet advertising market. The Commission believes that some Web sites may misuse consumer information and that consumers cannot tell which Web sites do and which do not. In recommending regulation of Internet privacy, the FTC cites surveys that suggest that "privacy concerns may have resulted in as much as $2.8 billion in lost online retail sales, while another suggests potential losses of up to $18 billion by 2002..." [81] If true – that is, if consumers refrain from purchasing over the Web because of insufficient information about Web sites' privacy policies – this would be an asymmetric-information market failure.

In the next chapter, we discuss in detail market responses to privacy concerns that are solving any asymmetric information problems that may exist. In the interim, the following points should be noted concerning market responses to asymmetric information. First, there is strong evidence that firms that violate privacy expectations lose business if those violations go uncorrected. Second, Web sites have a competitive incentive to inform consumers about the privacy protections they provide and, in fact, are doing so. Mechanisms, such as certifications by third parties, are now available for firms to differentiate themselves and to credibly commit to protecting privacy. Once such mechanisms exist and consumers learn about them, the conditions for market failure no longer hold. Finally, new technologies are enabling consumers to tailor the amount of information transferred to their own preferences.

It is also important to note that, in general, increased use of personal information solves, rather than exacerbates, asymmetric information problems, because it helps create informational symmetry. For example, information used in credit rating gives the lender additional information about the potential borrower, so that information becomes less asymmetric.

Moreover, the ability to sell for advertising or marketing purposes information initially collected for credit or insurance rating purposes increases the value of that information. Thus, the markets for advertising and

[81] Federal Trade Commission, *Privacy Online: Fair Information Practices in the Electronic Marketplace*, May, 2000, available online at http://www.ftc.gov/os/2000/05/index.htm#22, p. 2, footnotes omitted.

marketing information generate increased information in markets that might truly be susceptible to asymmetric information – i.e., credit and insurance markets. This contributes to reducing, rather than increasing, market failures due to asymmetric information. Regulation that reduced the amount of information being collected would have the opposite effect.

EXTERNALITIES

Market efficiency requires that buyers and sellers internalize the benefits and costs of transactions. If a transaction imposes costs on, or provides benefits to, third parties, then a market failure may exist. The classic example of a negative externality – a situation in which costs are imposed on third parties – is pollution. The parties do not consider the full costs of an activity when deciding how much to undertake – because they do not incorporate the cost of pollution – and therefore the polluting activity will be undertaken at too high a level. Government policies such as taxes or quotas are a solution to this problem.

Similarly, positive externalities exist when an activity providesbenefits to third parties. In this case, because the decision makers do not consider the full benefits, they will undertake too little of the activity. A solution is government subsidy of the activity in order to induce parties to undertake the activity at a higher level.

Since information is a good that can be used repeatedly once it is produced, many activities involving information create positive externalities. This partially explains government subsidy of information-related activities, such as education and scientific research.

The discussion in Chapter 2 suggests there are positive externalities associated with the commercial information market. First, once information is produced, the marginal cost of additional uses is close to zero. Credit bureaus, insurance companies, advertisers, direct sellers, and others can all use the same information at little additional cost and without any reduction in its value. (Of course, use of the same information by direct competitors may well reduce its value to each of them). Indeed, the various information users cooperate in generating this information because they all find it useful. This multiple use increases value and is a real benefit of an information economy.

If the information used in making credit decisions (or insurance decisions) can be sold to third parties, such as advertisers, this increases the value of the information and therefore creates incentives to increase the amount of information available. This in turn means that credit and insurance markets will function better. Thus, the market for advertising and marketing information creates an external benefit in the market for credit and insurance information, and conversely. With more information, all these markets work better. If regulation reduces the amount of information, then it increases possibilities of market failure.

Positive Externalities from Better Targeting

There are also positive externalities within the advertising information market itself. Advertisers use information in order to better target their messages – that is, deliver them to those consumers most likely to be interested in the advertiser's products. Recipients of unwanted messages hold the particular seller accountable, in the sense of being more likely to disregard future messages from this seller.[82] But, there is also a spillover effect on other sellers. Every time an unwanted message is received, the expected value of future messages from all senders is reduced and consumers become more likely to simply ignore all messages. Thus, the ability of any individual advertiser to obtain the information necessary to deliver targeted advertising benefits other advertisers – which is a positive externality.

One e-mail service provider has made this externality argument explicit. In discussing the optimal rate for e-mail, this firm recommends no more than one message per week, but also indicates that too many messages from any sender can be harmful and that "one of our biggest challenges" is that "we can't control what other people are doing out there."[83] The issue of excess messages leading consumers to ignore all messages is called "marketing clutter."[84]

Moreover, much of the information that sellers gather to target consumers is used for statistical modeling. For example, through statistical modeling, marketers may discover that purchasers of scientific books are more likely than average to purchase imported automobiles. Information of this sort needs to be collected from large numbers of consumers to be useful for statistical purposes. Information about a single consumer is only one data point for performing the statistical calculation. If sellers have information

[82] Julius L. Loeser, "Some Practical and Theoretical Thoughts about Privacy and Banking," in Competitive Enterprise Institute, *The Future of Financial Privacy*, Washington, 2000.
[83] Thomas E. Weber, "Why Companies Are So Eager To Get Your E-Mail Address," *The Wall Street Journal*, February 12, 2001.
[84] Ross D. Petty, "Marketing Without Consent: Consumer Choice and Costs, Privacy and Public Policy," 19 *Journal of Public Policy and Marketing*, Spring, 2000, 42-53, at 43.

about fewer consumers, it is more difficult to measure such relationships, and therefore more difficult to deliver targeted advertising. Because of this, provision of personal data by any individual has benefits that go beyond the services that particular individual might enjoy – i.e., it enables other individuals to be better targeted. This is yet another example of a positive externality. Thus, restricting the flow of data from some individuals would have an adverse effect on others.

Crime as an Externality

Finally, the possibility of crime – for example, credit card fraud or identity theft – creates an externality because these activities impose costs on individuals. However, many of these costs have been internalized. For example, most of the costs of credit card fraud are in the first instance borne by the credit card companies, since there is a $50 cap on direct losses to consumers. This gives the credit card companies a strong incentive to establish mechanisms, through their networks, to deter credit card fraud. Since the companies are in a better position to do this than individual credit card holders, this arrangement is probably efficient. Consumers, in effect, are provided with an insurance policy by their credit card company, but also retain a significant incentive to avoid losses because of the $50 loss and the general inconvenience of having to deal with loss of a credit card. Of course, the government also has an important role to play, in enforcing laws against fraud and theft.

MARKET POWER

The third major cause of market failure – market power – occurs when there are sufficiently few sellers in the market that they have some control over prices. When this occurs, prices are higher, and quantities produced lower, than they would be under competition, and some potentially value-increasing exchanges do not occur. The market for advertising information does not appear to be subject to market power. Some regulatory proposals do create competitive concerns, however, because they would make entry more difficult. This is discussed in more detail in Chapter 6.

CONSUMER HARM?

Even though commercial information markets do not appear to be subject to market failure, there could still be some incidence of consumer harm. It is, however, striking that, with all the concerns about privacy and the proposals to regulate, there is no real evidence of harm from the legitimate commercial use of information. In an economy with over 280 million persons, many of whom transact frequently on the Internet, unusual events will occur. But, if adverse events resulting from the commercial use of information have occurred, they have not made it into the public domain. There are several pieces of data consistent with this argument:

- The Federal Trade Commission's major report recommending new regulation of information on the Internet provides no evidence of actual harm to consumers.[85] The report discusses privacy concerns and privacy policies but presents no evidence about harm from data misuse to any actual consumer, and, indeed, the FTC did not find any such evidence.[86]

- The National Association of Attorneys General (NAAG) recently circulated a draft policy statement on privacy that includes examples of "troubling information sharing practices." [87] None of the examples, however, are examples of harm to consumers from the commercial use of information. Two of the examples concern what the NAAG considers improper sharing of information – with, however, no evidence of resulting harm. The other two are both "alleged" and concern misuse of consumer credit card numbers – which is already illegal. One of the episodes resulted in a fine of $13 million. Had there been an incident that caused real consumer harm and could not be addressed under

[85] Federal Trade Commission, *Privacy Online: Fair Information Practices in the Electronic Marketplace*, May, 2000, available online at http://www.ftc.gov/os/2000/05/index.htm#22.
[86] FTC Advisory Commission Report, separate statement of Stewart Baker, p. 40.
[87] Draft Memo, December 11, 2000, National Association of Attorneys General.

current law, the NAAG, as the association of the principal state law enforcement officers, presumably would be in a position to know.

- In a year-end summary for 2000 dealing with privacy issues, CNET, a leading "new economy" news source, indicated that there were no mishaps involving commercial use of personal information in 2000: "Despite the fears and concerns, there were no publicized horror stories that resulted from a privacy invasion."[88]

- In a magazine for electrical and electronics engineers, an academic expert (who, despite the lack of evidence, appears to believe there is a problem) is quoted: "But there has yet to be a single earth-shaking event that makes it obvious to everyone just how powerful and intrusive the technology can be."[89]

- Congressional hearings typically have witnesses who are victims of whatever issue is being discussed. For example, at hearings on Health Maintenance Organizations, consumers who have been denied coverage for some medical procedure might testify. But at privacy hearings, witnesses are typically college professors, industry representatives, privacy advocates, and representatives of think tanks and other organizations. There are no harmed individuals testifying at these hearings.[90] Moreover, the

[88] Patricia Jacobus, "Privacy heats up but doesn't boil over," CNET News, December 22, 2000, available online at http://news.cnet.com/news/0-1005-200-4238135 .html? tag=st. cn.sr.ne.1, visited December 25, 2000.

[89] Jean Kumagai, "Technology Priorities for the White House," *IEEE* [Institute of Electrical and Electronics Engineers] *Spectrum*, September 2000, 52-56, quoting Jon Peha of Carnegie Mellon University, at 55.

[90] Victims of identity theft do testify on occasion. However, this is not a problem associated with the legitimate use of information, nor is it particularly an Internet related problem.

> testimony provides no examples of such individuals.[91]

> The press generally writes stories using substantial anecdotal evidence about particular individuals in order to motivate the story, but one can look far and wide without finding one that actually points to a specific individual who has been harmed by legitimate use of data collected for advertising or marketing purposes. Rather, the stories contain examples of individuals who are "concerned" about privacy, without having experienced any harm, or individuals who have been harmed, but by activities that are not connected with the use of information for marketing and advertising purposes. For example, a *Business Week* story provided an anecdote about a consumer (described as shopping intensely online) who, upon learning about the use of information by Web sites where she shopped "was angry at first, then confused." But the story documents no other harm that befell this consumer.[92] A *New York Times Magazine* article discussed Monica Lewinsky's e-mails and various archives kept by chat rooms, and employer monitoring of e-mail and surfing.[93] While these examples may involve harm, they have nothing to do with the commercial use of information for marketing or advertising purposes.

Of course, lack of evidence of harm is not proof that there is no harm. Nonetheless, given the attention paid to this issue, the inability to find people who have been injured shows that such injury is, at a minimum, rare.

[91] This is in contrast to problems like identity theft, where there are real victims, as a recent *Wall Street Journal* article indicates: "Unlike with most privacy legislation, sponsors of identity-theft bills could make their cases with real examples of wrecked credit ratings and ruined finances. State Sen. Margarita Prentice, a Seattle Democrat who sponsored the Washington bill, says much of the commercial privacy debate – focusing on such issues as whether consumers are harmed when their shopping habits are tracked – is 'too theoretical.' But with identity theft, she says, 'there are genuine victims.'" Robert Gavin, "Lawmakers Crack Down on Identity-Theft Issues," *The Wall Street Journal*, May 16, 2001.

[92] *Business Week*, April 5, 1999, "Privacy: The Internet Wants Your Personal Info. What's In It For You?" In a story one year later, *Business Week* still advocated privacy regulation, but still had no evidence – even anecdotal – for its position: "It's Time for Rules in Wonderland," March 20, 2000.

[93] Jeffrey Rosen, "The Eroded Self," *New York Times Magazine*, April 30, 2000, p. 46.

SUMMARY

The market for commercial information used by the marketing and advertising sectors appears to be working well. There does not appear to be evidence of market failure, or even of consumer harm, from the legal use of information for these purposes. This is striking, given the number of individuals and transactions involved.

In general, markets work better with more information, not less. The analysis presented here and in the next chapter indicates that consumers are informed about the privacy policies of Web sites. There does not appear to be an information asymmetry problem with respect to this type of information. Consumers are in a position to make informed decisions about the Web sites they visit.

Limiting the flow of commercial information could, however, create information asymmetries. This could be particularly costly in the credit market, for example, which is dependent on much of the same information used by advertisers.

There are significant positive externalities associated with the commercial information market. Thus, to the extent a market failure exists, it results in less information than is socially optimal. Further reductions in information availability due to regulatory restrictions would exacerbate this problem.

Chapter 4

Market Reactions to Consumer Concerns

Although the Internet is still very young – commercial use of the medium did not begin to develop until the mid-1990s – market forces are moving rapidly to provide the privacy desired by consumers, in part by eliminating problems of asymmetric information. Perhaps most importantly, firms that do business on the Internet are discovering that there are substantial "reputation" costs associated with adopting information practices that are inconsistent with consumers' expectations. Firms respond by modifying their practices and by avoiding practices that may not be greeted favorably by their customers.

Firms are also adopting a variety of voluntary standards that involve third-party certification of privacy policies. This provides information to consumers and assures them that minimum standards are being met. Firms have an incentive to advertise their privacy policies to attract customers and the evidence suggests they are beginning to do so. In addition, new technologies are being developed that permit consumers directly to control the flow of their personal information to Web sites.

Finally, given full information about privacy policies, the market is adapting to provide an array of choices. This entails lower prices for those willing to trade their information and higher prices for those unwilling to do so.

REPUTATION EFFECTS

Firms that use personal information in ways their customers do not like face the prospect that their customers will stop doing business with them. Such reputation effects are powerful, and research has shown that firms ignore them at their peril. Firms that take actions that the markets perceive as harming their reputation with consumers invariably suffer a substantial loss in value.[94]

On the positive side, consumers reward firms that follow policies that they do like. Firms understand this and are beginning to advertise their privacy policies and use them as a competitive tool in the marketplace.[95]

Because communication on the Internet is relatively easy, consumers (or at least a sufficient number of them to cause a market reaction) can quickly learn about what they perceive as misdeeds by a firm. There are a number of prominent examples of the market disciplining firms that have violated consumers' preferences with respect to privacy:

- When Amazon appeared to be using personal information for "dynamic pricing" (what economists call price discrimination) consumers learned about it and many became irate.[96] Amazon responded quickly and promised not to adopt this practice. (We note, however, that economists do not consider this practice detrimental to consumers.[97] In fact, it can provide benefits in the form of products that otherwise would not be available.[98])

- America Online (in 1997) cancelled plans to sell telephone numbers of its subscribers to telemarketers in response to angry reactions from subscribers.[99]

[94] The theoretical argument was made by Benjamin Klein and Keith B. Leffler, "The Role of Market Forces in Assuring Contractual Performance," 89 *Journal of Political Economy* 615, 1981. For a summary of the empirical literature, see Kari Jones and Paul H. Rubin, "Effects of Harmful Environmental Events on the Reputations of Firms," *Advances in Financial Economics*, 2001, edited by Mark Hirschey, Kose John and Anil K Makhija, available online at http://papers.ssrn.com/paper.taf?ABSTRACT_ID=158849.

[95] Alec Klein and Shannon Henry, "On Reflection, a Puzzling Ad Campaign," *The Washington Post*, March 1, 2001, p. E1.

[96] David Streitfeld, "On the Web, Price Tags Blur," *Washington Post*, September 27, 2000. Amazon denies that it was engaged in dynamic pricing or price discrimination.

[97] Paul Krugman, "What Price Fairness?," *New York Times* October 4, 2000;

[98] Carl Shapiro and Hal Varian, *Information Rules: A Strategic Guide to the Network Economy*, Cambridge, MA: Harvard Business School Press, Chapters 2 and 3.

[99] This and the following two examples are from Jessica Litman, "Information Privacy/Information Property," 52 *Stanford Law Review*, 1283-1313, May, 2000, at 1305-6.

- CVS pharmacy (in 1998), in response to consumer dissatisfaction, called off plans for another company to contact consumers who failed to refill prescriptions.

- RealNetworks (in 1999) changed its software when it was learned that its product, RealJukebox, collected information on users' habits.

- Yahoo! eliminated the reverse telephone number search from its site in response to consumer unhappiness.[100]

- Lotus cancelled plans to sell data about 120 million citizens.

- Lexis-Nexis also cancelled plans to sell information about millions of persons.

- More recently, a firm called N2H2, which makes filtering software, has stopped selling information about Web sites visited by students because many felt that such sales were improper, even though all information was used anonymously.[101]

- Finally, there is the well-known story of DoubleClick's cancelled plan to link online and personally identifiable information though its acquisition of Abacus Direct.[102] There is some evidence that this episode led to a substantial fall in the price of the stock of DoubleClick, which would

[100] This and the following two examples are from Daniel J. Solove, "Privacy and Power: Computer Databases and Metaphors for Information Privacy," p. 27, available online through SSRN.Com, 56-57.
[101] Associated Press, "Internet Co. Drops Data Selling Plan," Feb. 22, 2001. Note that the plan did not sell personally identifiable information.
[102] Discussed at numerous places. See for example Diane Anderson and Keith Perine, "Marketing the Double Click Way," *The Standard*, March 13, 2000.

- be consistent with the empirical evidence for other such events.[103]

These examples illustrate that when businesses use information in a way that consumers do not like, consumers quickly learn about this use, and the firms are forced to stop. Such reputational penalties are among the strongest protections available to consumers. In fact, the principal asset that online marketers have is their reputation with consumers, and any use of information in a way that reduces the value of those reputations is counterproductive for the firm.

The fact that consumers become informed quickly and are able to make their displeasure known in the marketplace suggests that asymmetric information is not a problem in these markets. It has been argued that consumers do not have adequate incentives to learn about the privacy policies of Web sites, and therefore the Web sites will not have adequate incentives to provide privacy policies that are optimal.[104] It may be true that some consumers do not take the time to learn about Web sites' privacy policies, but, as the examples described above demonstrate, there are a sufficient number of consumers who are informed to produce a strong market reaction, if they find the policies not to their liking. Due to consumer interest, the press and privacy advocates are also monitoring these markets. The reaction occurs *ex post* – after a violation – rather than *ex ante* – before a transaction. This is sufficient to send a powerful message to firms that they will incur losses if their information management policies are not to their customers' liking. Firms, therefore, have a strong incentive to avoid undertaking policies that run the risk of offending their customers.

Firms are also taking positive steps to protect their reputations. Companies like IBM, Microsoft, Disney, Intel, Compaq, Novell, Procter & Gamble, and American Express do not advertise on Web sites that do not have privacy policies.[105] Firms are hiring Chief Privacy Officers ("CPOs"),

[103] Will Roger and Gregg Farell, "Investors Dump DoubleClick," *USA Today*, February 17, 2000. However, this fall in value was short lived: "Double Click Bounces Back on report," *USA Today*, February 25, 2000. We note that FTC investigation of DoubleClick found they had not violated any FTC requirements.
[104] Peter Swire, "Markets, Self-Regulation, and Government Enforcement in the Protection of Personal Information," in Privacy and Self-Regulation in the Information Age, U. S. Department of Commerce, Washington, DC, 1997, http://www.ntia.doc.gov/reports/privacy /selfreg1.htm.
[105] "It's Time for Rules in Wonderland," *Business Week*, March 20, 2000; "Towards Digital eQuality – The Second Annual Report of the US Government's Work Group On Electronic Commerce," December, 1999.

and giving them substantial power and discretion in setting company policies.[106] Alan Westin, a well-known privacy advocate and expert, offers a training course for this position in association with his organization, Privacy & American Business.[107] There are now about 100 CPOs, and it is estimated that there will be 500 by the end of next year. They are paid about $120,000-$175,000 per year.[108]

VOLUNTARY STANDARDS

Voluntary standards – "seal" programs – defined, explained and enforced by third parties or by consortia of Web operators, are an important mechanism for providing information to consumers about Web sites' information policies and responding to consumers' privacy concerns. The fact that these standards are enforced by known third parties provides consumers with additional assurances. And, of course, once a claim is made by a Web page (whether associated with a third-party program or not), it is enforceable by the FTC.

The following voluntary third-party certification programs are now operating:

- TRUSTe is a voluntary association that audits Web sites to ensure compliance with specified privacy policies.[109] It has 1,570 member firms. When a Web site joins TRUSTe, a link is put on its Web site connecting the visitor to the site's privacy policy.[110]

- The Better Business Bureau has a certifying program, BBBOnLine, that performs similar functions.[111]

[106] IBM has been running a series of newspaper advertisements featuring a full-page picture of its Chief Privacy Officer, Harriet Pearson. See, for example, *The Wall Street Journal* June 6, 2001, p. A14.
[107] http://www.pandab.org/.,visited November 13, 2000.
[108] Kemba J. Dunham, "The Jungle: Focus on Recruitment, Pay and Getting Ahead: A New Playing Field," *The Wall Street Journal,* March 20, 2001.
[109] Website: http://www.truste.org.
[110] Found at http://www.truste.org/users/users_lookup.html visited December 22, 2000.
[111] http://www.bbbonline.org/.

- The Direct Marketing Association has various voluntary standards in place, including a method consumers can use to have their names removed from e-mail lists. In addition, members of the DMA are required (as a condition of membership) to meet certain standards regarding privacy on the Web.[112]

- Auditing firms, such as PriceWaterhouseCoopers and Ernst & Young, perform privacy audits and place a box on a Web site indicating that the site conforms to their stated privacy policies.[113]

TECHNOLOGIES OF CHOICE

A number of technologies have become available that allow consumers to control their information. Basic browsers now allow some customization with little effort. For example, both Netscape and Microsoft allow users a variety of options with respect to cookies.

There are other options as well. One site provides the consumer the opportunity to download approximately forty programs that allow the consumer to refuse cookies, or to easily delete them after they are received.[114] DoubleClick allows consumers to refuse all of its cookies.[115]

Internet users can surf anonymously using services such as Anonymizer.com, IDZap.Com, iPrivacy.com, SafeWeb and SilentSurf.com. These services offer different levels of privacy, depending on the consumer's preferences and willingness to bear inconvenience costs. In addition, American Express now offers a "one-time" credit card number, good only for one purchase, designed for Internet use. Since Web sites selling products to consumers using this card never have access to information about the consumer, privacy is protected. American Express also sells a prepaid card through 7-Eleven stores. While these are mainly aimed at individuals with poor credit ratings, they can also be used for anonymous purchases, on the

[112] http://www.the-dma.org.

[113] Bob Tedeschi "Sellers Hire Auditors to Verify Privacy Policies and Increase Trust," *New York Times*, September 18, 2000.

[114] Downloaded on October 25, 2000 from ZDNet Downloads http://www.zdnet.com/downloads/, a popular source for software, using a search for "cookie".

[115] http://www.doubleclick.net:80/us/corporate/privacy/opt-out.asp?asp_object_1=&.

Internet or elsewhere.[116] Consumers concerned about privacy are able to use any of these services, some free, to protect their information online.[117]

Simpler methods of self-protection are also available. In one non-random survey, about 50 percent of respondents indicated that they had provided false information when registering at a Web site – an easy form of do-it-yourself privacy.[118] This survey also found that the most common reason for not registering at a Web site is that the terms and conditions of the use of information are not clearly specified, or that revealing the requested information is not worth registering and being able to access the site.[119] One study found that as consumers are more concerned about privacy, they undertake more actions to protect privacy, such as refusing to register at a Web site, providing false or incomplete information, and asking to have names removed from lists.[120]

Technologies on the horizon will provide other solutions. "Trusted systems" are envisioned as computer protections that limit the way in which data can be copied. These technologies are being developed to protect intellectual property such as music, movies and books, but it may be possible to adapt them to protect consumer information as well.[121]

THE PLATFORM FOR PRIVACY PREFERENCES (P3P)

Some firms are large enough to be able to internalize production of what otherwise would be a public good. For example, the World Wide Web Consortium (W3C),[122] with about 500 members, including the largest players on the Internet, such as Microsoft, America Online and Cisco, is in the process of drafting a major private privacy protocol, the Platform for Privacy

[116] Declan McCullagh, "Prepaid Phones and Privacy, Too," *Wired*, March 14, 2001.
[117] Some of these are discussed in Don Clark, "Privacy: You Have No Secrets," *The Wall Street Journal*, October 23, 2000 and Lorrie Faith Cranor, "Agents of Choice: Tools That Facilitate Notice and Choice about Web Site Data Practices," available online from http://www.research.att.com/~lorrie/#publications.
[118] GVU's 7th WWW User Survey, http://www.gvu.gatech.edu/gvu/user_surveys/survey-1998-04/
[119] GVU's 7th WWW User Survey, http://www.gvu.gatech.edu/gvu/user_surveys/survey-1997-04/
[120] Kim Bartel Sheehan and Mariea Grubbs Hoy, "Flaming, Complaining, Abstaining: How online users respond to privacy concerns," 37 *Journal of Advertising*, 1999.
[121] Jonathan Zittrain, "What the Publisher Can Teach the Patient: Intellectual Property and Privacy in an Era of Trusted Privication, 52 *Stanford Law Review* 1201-1250, May 2000.
[122] For the W3C homepage, see http://www.w3.org. For the list of members, see http://www.w3.org/Consortium/Member/List, visited December 22, 2000, when there were 488 members.

Preferences, P3P.[123] If P3P is successful, it will provide standardized information in machine-readable form about each Web site's privacy policy. Individuals will then be able to configure their own browsers to interact with the Web site. Microsoft will soon begin incorporating P3P standards in its software.[124] They will also be available as a downloadable plug-in.[125]

The P3P software will solve one side of the "chicken and egg" problem. Since the software will be available to consumers, Web sites will have a ready-made audience if they install the other side of the package. P3P also offers the possibility of increased negotiation and customization of privacy policies.[126]

PRICE EFFECTS

Finally, different Web sites have different information policies and this is reflected in the prices they charge and/or the services they provide. A simple theoretical example illustrates the point.

Consider two competing Web sites both selling a product – say, computer chips. Site CP has a strong privacy protection policy. CNP has no privacy policy, and makes use of the information provided by consumers for other purposes. Then CNP will sell chips cheaper than will CP, because it earns revenue from the sale of information received from consumers and so can charge a lower price for chips and still make a profit. Consumers who prefer more privacy will deal with CP, and consumers who prefer lower prices will deal with CNP.

This, in fact, is what we observe on the Internet, although the lower prices may take the form of free or discounted services. As discussed in Chapter 2, portal sites such as Yahoo! provide a whole range of free services in exchange for information from their customers. There are numerous other examples of companies that provide discounts and benefits to consumers willing to share their information. AllAdvantage.com pays consumers to monitor their browsing, and some consumers (presumably those less concerned with privacy issues) are apparently willing to join this program.[127] Dash.com provides discounts to consumers who allow monitoring. Thus,

[123] http://www.w3.org/P3P/.
[124] "New Tools to Help Web Surfers Protect Privacy," Associated Press, June 22, 2000.
[125] Elizabeth Weise, "Privacy plug-in will ask: 'Do you want to go there?'," *USA Today*, July 11, 2000.
[126] See, for example, Lawrence Lessig, "The Architecture of Privacy," 1998, Online, http://cyber.law.harvard.edu/works/lessig/architecture_priv.pdf.
[127] http://www.alladvantage.com/home.asp?refid=

consumers have different preferences regarding Internet privacy, which the markets are now satisfying.

SUMMARY

The market is responding to concerns about privacy in a variety of ways. First, there are strong market pressures on Web site proprietors to provide levels of privacy that consumers desire. Web sites that have violated consumers' expectations suffer a loss in reputation that is costly in terms of lost sales and market value. Second, the market is providing voluntary standards, certified and enforced by reputable third parties. Third, the market is developing new technologies that can be used by individual consumers to better control the flow of their information on the Web. And, finally, the market is providing consumers with a range of options, whereby they can exchange their information for valuable services.

Chapter 5

Polls, Policy and the FTC

The last two chapters demonstrated that markets for commercial information are responding to consumers concerns and preferences with respect to privacy. There appears to be no evidence of market failure, or even of harm to consumers. There are, nonetheless, proposals for regulation, most prominently from the Federal Trade Commission.[128] The National Association of Attorneys General (NAAG) has also proposed greater government involvement.[129] In this chapter, we consider alternative rationales for expanded regulation, which seem largely to be based on public opinion and other survey data.

DO CONSUMERS WANT REGULATION?

Consumer surveys need to be interpreted with care, because the answers are heavily dependent on the way the questions are asked, and because it is easy for individuals to say they want more of a particular good (e.g., privacy) when not being made aware of potential costs. Even so, surveys present an ambiguous picture concerning whether or not consumers want more regulation in this area. Consider the conflicting results of various surveys presented in Table 1.

[128] Federal Trade Commission, *Privacy Online: Fair Information Practices in the Electronic Marketplace*, May, 2000, available online at http://www.ftc.gov/os/2000/05/index.htm#22. The new FTC Chairman has indicated that he wants to review these recommendations.
[129] Draft Memo, December 11, 2000, National Association of Attorneys General.

Table 1: What Consumers Think About Privacy Regulation

SURVEY	***QUESTION***	***RESPONSE***
The Wall Street Journal-Harris Interactive Poll, "Exposure in Cyberspace," 2001	Percentage of respondents who think the government should regulate privacy online more strictly.	38%
Harris Interactive: Business Week/Harris Poll: A Growing Threat, 2000	Percentage of respondents who believe the government should pass laws regulating the collection of personal information.	57%
Louis Harris and Associates, Inc.: E-Commerce and Privacy: What Net Users Want, 2000	Percentage of respondents that buy products and services online and believe that business incentives will be enough to encourage companies to adopt good privacy standards without.	51%
Pew Internet and American Life Project: Trust and Privacy Online, 2000	Percentage of respondents who believe that businesses should set the rules governing how tracking of consumers online is done.	18%
Pew Internet and American Life Project: Trust and Privacy Online, 2000	Percentage of respondents who believe that internet users should set the rules governing how tracking of consumers online is done.	50%
Pew Internet and American Life Project: Trust and Privacy Online, 2000	Percentage of respondents who believe that government should set the rules governing how tracking of consumers online is done.	24%
AT&T Labs-Research: Beyond Concern: Understanding Net Users' Attitudes about Online Privacy, 1999	Percentage who said they would be more willing to provide information if there were a law that prevented the site from using the information for any purpose other than processing the request.	48%

Table 1: What Consumers Think About Privacy Regulation (Cont'd)

SURVEY	*QUESTION*	*RESPONSE*
AT&T Labs-Research: Beyond Concern: Understanding Net Users' Attitudes about Online Privacy, 1999	Percentage who said they would be more willing to provide information if the site had a privacy policy and a seal of approval from a well-known organization such as the BBB.	58%
AT&T Labs-Research: Beyond Concern: Understanding Net Users' Attitudes about Online Privacy, 1999	Percentage who said they would be more willing to provide information if the site had a privacy policy.	28%
@plan, "Internet Poll," 1999	Percentage of respondents who believe that the Internet industry can better protect consumers than the Federal Government.	64%
IBM Multi-National Consumer Privacy Survey, 1999	Percentage of U.S. respondents who somewhat or strongly agree with the statement "Most businesses handle the personal information they collect about customers in a proper and confidential way".	64%
IBM Multi-National Consumer Privacy Survey, 1999	Percentage of U.S. respondents who somewhat or strongly agree with the statement "Existing laws and organizational practices in the United States provide a reasonable level of consumer privacy protection today".	59%

Sources: @plan, "Internet Poll," November, 1999, summary available at www.webplan.com; AT&T Labs, "Beyond Concern: Understanding Net Users' Attitudes about Online Privacy". Available at http://www.research.att.com/library/trs/TRs/99/99.4/99.4.3/report.htm. April, 1999; Harris and Associates, "E-Commerce and Privacy: What Net Users Want;" Harris Interactive, "Business Week/Harris Poll: A Growing Threat." Available online at http://www.businessweek.com/2000/00_12/b3673010.htm; IBM, "Multi-National Consumer Privacy Survey". October, 1999; Louis Harris and Associates, Inc. and Dr. Alan F. Westin, "E-Commerce and Privacy: What Net Users Wan," press release.

While one study (the Harris Interactive study) suggests a slight majority in favor of regulation, other studies presented in Table 1 do not support that conclusion. Sixty percent of respondents to a recent *Wall Street Journal*-Harris Interactive poll do not support an expansion of regulation. This is consistent with a 1999 IBM survey in which a similar percentage of respondents thought businesses handled consumer information appropriately. The AT&T survey indicates that there is a significant demand for third-party certification (seal) programs. The Pew Internet and American Life study respondents indicated that they do not want either government or business to establish the rules governing how consumers are tracked online. This suggests a large role for technologies like P3P, which give consumers the power to protect their privacy.

Responses to questions about the need for regulation also need to be placed in the context of the public's understanding of how information is used on the Internet. As discussed in Chapter 2, commercial information is sold in blocks of 1,000 consumers with certain characteristics, and advertisers access consumers in this way. Individual consumers remain anonymous. Anyone with an interest in examining these lists would have to look for a long time to find the name of a particular individual. Indeed, in general, the sale of information involves transfers of data between computers.

This mechanism is not well understood by consumers, and even some commentators on the issue. Some of the support for regulation is probably attributable to not understanding how advertising information is actually processed and used.

For example, consider the following from a widely-cited article by Jerry Kang from the 1998 *Stanford Law Review*.[130]

> By contrast, in cyberspace, the exception becomes the norm: Every interaction is like the credit card purchase. The

Avavliable online at http://www.pandab.org/E-Commerce%20Exec.%20Summary.html. July, 2000; Pew /internet and American Life Project, "The Internet Life Report: Trust and Privacy Online: Why Americans want to rewrite the rules." August, 2000. More information available at www.pewinternet.org; The Wall Street Journal-Harris Interactive Poll, "Exposure in Cyberspace," The Wall Street, March 21, 2001.

[130] Jerry Kang, "Information Privacy In Cyberspace Transactions," 50 *Stanford Law Review* 1193, April 1998, at 1198-1199, footnotes omitted. As of December 26, 2000, Westlaw indicates that this article had been cited 100 times – a large number for a relatively new article. Moreover, it has been cited in many or most of the law review literature on privacy in cyberspace.

> best way to grasp this point is to take seriously, if only for a moment, the metaphor that cyberspace is an actual place, a computer-constructed world, a virtual reality. In this alternate universe, you are invisibly stamped with a bar code as soon as you venture outside your home. There are entities called "road providers," who supply the streets and ground you walk on, who track precisely where, when, and how fast you traverse the lands, in order to charge you for your wear on the infrastructure. As soon as you enter the cyber-mall's domain, the mall begins to track you through invisible scanners focused on your bar code. It automatically records which stores you visit, which windows you browse, in which order, and for how long. The specific stores collect even more detailed data when you enter their domain. For example, the cyber-bookstore notes which magazines you skimmed, recording which pages you have seen and for how long, and notes the pattern, if any, of your browsing. It notes that you picked up briefly a health magazine featuring an article on St. John's Wort, read for seven minutes a newsweekly detailing a politician's sex scandal, and flipped ever-so-quickly through a tabloid claiming that Elvis lives. Of course, whenever any item is actually purchased, the store, as well as the credit, debit, or virtual cash company that provides payment through cyberspace, takes careful notes of what you bought – in this case, a silk scarf, red, expensive.

Similarly, the following quote from an article by Jeff Sovern:[131]

> The information available on consumers is striking. For example, you can buy lists of people who have bought skimpy swimwear; college students sorted by major, class year, and tuition payment; millionaires and their neighbors; people who have lost loved ones; men who have bought fashion underwear; women who have bought wigs; callers to a 900-number national dating service; rocket scientists; children who have subscribed to magazines or have sent in

[131] Jeff Sovern, "Opting In, Opting Out, Or No Options At All: The Fight for Control of Personal Information," 74 *Washington Law Review* 1033, October, 1999. at 1034, footnotes omitted.

> rebate forms included with toys; people who have had their urine tested; medical malpractice plaintiffs; workers' compensation claimants; people who have been arrested; impotent middle-aged men; epileptics; people with bladder-control problems; buyers of hair removal products or tooth whiteners; people with bleeding gums; high-risk gamblers; people who have been rejected for bank cards; and tenants who have sued landlords. There are lists based on ethnicity, political opinions, and sexual orientation.

Because it is possible to buy lists of information by category does not mean that it is possible to easily obtain information about individuals in the categories. The lists are available because firms with products to sell are interested in buying lists of consumers characterized by criteria such as those mentioned above. As discussed in Chapter 2, businesses and advertisers do not look for a particular consumer and attempt to figure out what that consumer wants to buy, or whether the consumer has bought skimpy swimwear or fashion underwear. Rather, the process is the opposite: a seller has products to sell, and tries to find consumers who are more likely than average to buy those products.

Survey data from consumers is consistent with the fact that there is confusion about the way in which information is collected and used.[132] In one survey, a question asked:

> Businesses marketing goods and services directly to consumers are now able to buy from mailing list-making companies information about your consumer characteristics – such as your income level, residential area, and credit card use – and use such information to offer goods and services to you. Do you feel this is a good or bad thing?

Sixty-nine percent of consumers felt it was a bad thing. A second question in the same survey asked:

> Increasingly, companies are marketing goods and services directly to people by mail. Some reasons for this trend are that many people have less time to shop or they prefer to

[132] From the Equifax survey, cited in Jeff Sovern, "Opting In, Opting Out, Or No Options At All: The Fight for Control of Personal Information," 74 *Washington Law Review* 1033, October, 1999. p. 1061 and footnotes.

> make shopping decisions at home. Also, companies are trying to reduce their costs of advertising and selling in stores, and they find direct marketing can reduce their expenses and their product prices. Companies try to learn which individuals and households would be the most likely buyers of their products or service. They buy names and addresses of people in certain age groups, estimated income groups, and residential areas with certain shopping patterns so they can mail information to the people they think will be most interested in what they are selling. Do you find this practice acceptable or unacceptable?

When asked in this way, two-thirds found the practice acceptable.[133]

SURVEYS AS A BASIS FOR PUBLIC POLICY

The FTC report, which argues in favor of greater regulation, bases its recommendations primarily on survey data. First, the report cites an Odyssey survey in which 82 percent of households favored regulation. Odyssey's methodology was, however, biased in favor of obtaining a positive result. The survey provided two positive answers – "strongly agree that government should regulate" and "somewhat agree that government should regulate" – but only one strongly negative answer. No answer was provided for consumers who weakly disagree that the government should regulate. The report adds both categories of positive responses and concludes that an overwhelming majority of consumers want regulation.

Second, the FTC bases its recommendations on its own survey of privacy policies on Web sites. The FTC survey found that, in 2000, 88 percent of a random sample of Web sites and 100 percent of the most popular Web sites provided privacy disclosures. A 1998 survey had found that only 14 percent of a random sample of Web sites had at least one privacy disclosure. Thus, although the rate of voluntary provision of privacy information by Web sites increased dramatically, and the great majority of Web sites and all the most

[133] It is sometimes argued that the value of privacy is demonstrated by the willingness of many consumers to have unlisted phone numbers. But this is consistent with the argument here. A phone number is available to real people. It is quite consistent to value privacy with respect to what is known by humans more highly than privacy with respect to information held on computers. See Eli Noam, "Privacy and Self-Regulation: Markets for Electronic Privacy," in Privacy and Self-Regulation in the Information Age, U. S. Department of Commerce, Washington, 1997, http://www.ntia.doc.gov/reports/privacy/selfreg1.htm.

popular sites have explicit privacy disclosures, the FTC believes that progress has not been sufficient.

The FTC report concludes that "seals" programs – third-party certification of the type discussed in the previous chapter – are not prevalent enough, and that this failure of self-regulation implies a need for government regulation. However, the report ignores the possibility that sites that do not have seals (perhaps because they are costly) may be encouraged to adopt privacy policies by competitive pressures, or that there is a market niche for sites without any particular privacy policy. As discussed below, some consumers seem largely indifferent to the issue.[134]

The FTC believes, also on the basis of survey data, that lack of mandatory privacy requirements is inhibiting the development of the Internet. The FTC report sites two surveys, one by Jupiter Communications, the other by Forrester Research, alleged to show large losses in sales due to consumers' privacy concerns. Forrester estimates lost sales of $2.8 billion in 1999, and Jupiter estimates losses of up to $18 billion by 2002.

Since the FTC has not made either of these studies available, it is difficult to comment on their methodologies. But, they do not seem to provide a logical basis for regulation. If firms can attract sales by adopting superior privacy policies, they have every incentive to do so. Companies doing

[134] The FTC did not report on federal government Web sites, which appear not to have made as much progress – by the FTC's criteria – as private sites. See Government Accounting Office (GAO), "Information Security: Serious and Widespread Weakesses Persist at Federal Agencies; Government Accounting Office (GAO) (2000b), "Internet Privacy: Federal Agency Use of Cookies." GAO-01-147R, October 20 2000. Recall that the FTC found that 88 percent of a random sample of Web sites and 100 percent of the most visited sites had privacy disclosures; for the federal government, the corresponding figures are 84 percent of the random sample and 75 percent of the "high impact" Web sites. Indeed, questions have been raised about the government's sensitivity to several privacy issues. Two recent Government Accounting Office (GAO) reports have found that government Web sites routinely violate privacy principles. One of these studies finds that 12 of 65 government Web sites surveyed (18 percent) give cookies without notice; three of those give third-party cookies. See Government Accounting Office (GAO), "Internet Privacy: Federal Agency Use of Cookies," GAO-01-147R, October 20, 2000. Moreover, eight of 32 "high impact agencies, which handle the majority of the government's contact with the public" are among the agencies that do not even provide notice. In a somewhat different area, but also related to privacy, substantial questions have been raised about the privacy implications of the FBI's new Internet monitoring technologies. See "Reno Says Review Is Under Way on Internet 'Wiretapping'," Associated Press/ *New York Times*, July 14, 2000. Although a review has found that there are no difficulties with this program, others disagree: John Schwartz, "Review Released on Web Wiretap," *New York Times* November 22, 2000.

business on the Internet would seem to be in a better position than a government agency to know how to attract customers.[135]

The FTC study did not gather any information on or evaluate the specific privacy policies actually followed by the Web sites in its survey. It simply checked for explicit statements of policies and the policies were checked to make sure they contained some content.

Most importantly, the FTC report did not examine whether there was evidence of market failure or of harm to consumers.[136] Without such evidence, there is no case for increased regulatory involvement.

SUMMARY

Public opinion data are not a good substitute for public policy analysis.[137] Moreover, the opinion data that are available on privacy present an ambiguous picture concerning the public's desire for more regulation. The data appear to indicate that consumers want mechanisms that will empower

[135] We also note that the Supreme Court has rejected this argument in another context. In the case involving the Communications Decency Act (CDA) the government made the argument that the Internet would grow faster if there were regulation of pornography, an argument similar to that made with respect to protection of information. The opinion stated:

> In this court, though not in the District Court, the government asserts that – in addition to its interest in protecting children – its "equally significant" interest in fostering the growth of the Internet provides an independent basis for upholding the constitutionality of the CDA. The government apparently assumes that the unregulated availability of "indecent" and "patently offensive" material on the Internet is driving countless citizens away from the medium because of the risk of exposing themselves or their children to harmful material. We find this argument singularly unpersuasive. The dramatic expansion of this new marketplace of ideas contradicts the factual basis of this contention. The record demonstrates that the growth of the Internet has been and continues to be phenomenal. As a matter of constitutional tradition, in the absence of evidence to the contrary, we presume that governmental regulation of the content of speech is more likely to interfere with the free exchange of ideas than to encourage it. The interest in encouraging freedom of expression in a democratic society outweighs any theoretical but unproven benefit of censorship.

See Janet Reno, Attorney General of the United States, et al., Appellants v. American Civil Liberties Union et al., No. 96-511, Supreme Court of the United States, decided March 19, 1997, at 15-16.

[136] FTC Advisory Commission Report, separate statement of Stewart Baker, p. 40.

[137] For additional discussion of this issue, see Solveig Singleton and Jim Harper, "With a Grain of Salt: What Consumer Privacy Surveys Don't Tell Us," May 8, 2001, http://www.cei.org/PRReader.asp?ID=1469.

them to make informed choices rather than more government involvement. Much of the support for regulation that is exhibited in the surveys may stem from consumers not understanding that information used for advertising and marketing processes is used anonymously.

The FTC's basis for its regulatory recommendations is particularly weak. First, the agency cites a flawed public opinion survey that purports to show that a large majority of consumers favor regulation. Second, the FTC cites studies that find that businesses are losing large amounts of sales on the Web due to the absence of regulation. But these studies, which are also apparently based on surveys rather than analysis, are not available for outsiders to review. Finally, on the basis of its own survey, the FTC has concluded that not enough Web sites are providing privacy disclosures. This, despite the fact that, in 2000, all the most popular sites and 88 percent of all sites (up from 14 percent in 1998) do provide such disclosures. Clearly, the FTC has not undertaken the type of public policy analysis needed to determine if there is market failure or if the benefits of its proposals will outweigh their costs.

Chapter 6

The Effects of Regulation

The absence of serious market failure or consumer harm suggests that the potential benefits of new privacy regulations are very small. We, therefore, focus on the potential costs of such regulations. We first discuss some general problems that are likely to occur if this market is subject to more regulation and then examine in greater detail the proposals of the FTC and others.

INFLEXIBILITY OF REGULATION

Change is the normal state of affairs for the Internet, and for software and other products that interact with the Internet. The P3P release is P3P 1.0, indicating that, like software in general, the drafters expect the privacy policies embedded in the document to change over time. As discussed in Chapter 4, new technologies are continually being developed that permit consumers to better control their information.

Government, on the other hand, is slow to change. Laws passed by government do not come with release numbers because there is no expectation that they will be changed frequently or at all.

Any regulation of the online market for information at this time would likely freeze potentially important aspects of the Internet in their current state. Moreover, even if regulations were perfectly suited for today's environment, they would quickly become obsolete as the Internet changed.

Once an inefficient regulatory scheme is in place, it becomes very difficult to modify or eliminate. As we discuss in more detail later in this chapter, the FTC proposes to evaluate firms and Web sites according to whether or not they provide information according to specific categories. If it should turn out that other categories are better, the Internet would nonetheless be locked into the FTC's choices.

Government regulation is necessarily of the one-size-fits-all variety. This may be justified if all consumers have identical or similar preferences. When preferences differ, however, as they do with respect to privacy, some consumers are harmed. As one industry source puts it, "What's an invasion of privacy to one consumer is a great deal to another."[138]

The FTC itself acknowledges that consumers differ in their preferences with respect to Internet privacy: "According to one panelist, survey research consistently indicates that roughly one-quarter of the American public is 'intensely' concerned about privacy and that another quarter has little or no concern; the remaining fifty percent view this issue pragmatically..."[139]

These differences are documented in a survey on Internet privacy by AT&T.[140] Those most concerned about Internet privacy – those the AT&T report calls "privacy fundamentalists" – can already protect themselves using a variety of techniques discussed above. On the other hand, some consumers are so little concerned with privacy issues that they are willing to have all of their Web surfing monitored. Similarly, 78 percent would accept cookies to provide a customized service; 60 percent would accept cookies for customized advertising; and 44 percent would accept cookies that convey information to many web sites.

The AT&T report also finds that consumers have very different privacy preferences regarding different types of information. Consumers are less willing to provide Social Security and credit card numbers than other types of information.

This means that any useful privacy notice would have to be exceedingly complex – so complex that few people would be willing to read it. Moreover, different pages within the same site might require different policies, so virtually each mouse click would require reading a new notice. On the other hand, a protocol such as P3P could provide customized settings for each type of information and each potential use, based on consumers filling out a one-time form when configuring their browsers. Of course, some consumers would choose not to do so and would merely accept the defaults.

In the absence of regulation, some consumers choose to supply much information to Web sites; others choose to surf anonymously. Regulation would deny this choice to consumers. Even a notice requirement would deny

[138]Margaret Barnett, "The Profilers: Invisible Friends," *The Industry Standard*, March 13, 2000, p. 221.

[139] In its 1998 Report, Part II, at 2.

[140] Lorrie Faith Cranor, Joseph Reagle, and Mark S. Ackerman, "Beyond Concern: Understanding Net Users' Attitudes About Online Privacy," AT&T Labs-Research Technical Report TR 99.4.3, 1999 http://www.research.att.com/library/trs/TRs/99/99.4/.

consumers the option of using less expensive Web sites that did not offer notice.

CROWDING OUT PRIVATE MARKET RESPONSES

Regulation would "crowd out" private market responses of the sort discussed in Chapter 4 that are developing in response to privacy concerns. Such market responses include new technologies that empower consumers and voluntary third-party certification (seal) programs. The adoption of government-mandated policies reduces, and perhaps eliminates, incentives to adopt new technologies and voluntary programs, for which the government program is a substitute. In fact, such marketplace innovations may be inconsistent with regulations such as those contemplated by the FTC and others.

This is of particular concern, because there is evidence that the voluntary standards that companies have been adopting in the U.S. work better than mandatory standards imposed by the European Commission.[141] About twice as many U.S. sites (62 percent) as European sites (32 percent) have posted privacy policies. Moreover, although "opt-out" is required in Europe, only 20 percent of Web sites actually offer this option to consumers; in the U.S., 60 percent of sites offer this choice. Although all members of the EU now have data-privacy commissioners and agencies, these agencies seem unable to enforce privacy regulations. Thus, it appears that voluntary self-regulation provides more privacy than does mandatory government-imposed regulation. It may be that the existence of rules, even if the rules are not enforced, lulls consumers into believing that they have more protection than is actually the case, as predicted by Ira Magaziner.[142] This diminishes the incentives private firms would otherwise have to compete on the basis of their privacy practices.

FAIR INFORMATION PRACTICES

The FTC has proposed a regulatory regime that focuses on four fair information practices – Notice, Choice, Access, and Security. Some

[141] Ben Vickers, "Europe Lags Behind U.S. on Web Privacy: More American Firms Let Customers Guard Data, Study Finds," *The Wall Street Journal*, February 20, 2001.
[142] Ira C. Magaziner, "Creating a Framework for Global Electronic Commerce," Future Insight, The Progress and Freedom Foundation, July, 1999.

combination of these elements is incorporated in virtually all the regulatory proposals that are under consideration:[143]

- Notice. Web sites would be required to provide consumers clear and conspicuous notice of their information practices, including what information they collect, how they collect it (e.g., directly or through non-obvious means such as cookies), how they use it, how they provide Choice, Access, and Security to consumers, whether they disclose the information collected to other entities, and whether other entities are collecting information through the site.

- Choice. Web sites would be required to offer consumers choices as to how their personal identifying information is used beyond the use for which the information was provided (e.g., to consummate a transaction). Such choice would encompass both internal secondary uses (such as marketing back to consumers) and external secondary uses (such as disclosing data to other entities).

- Access. Web sites would be required to offer consumers reasonable access to the information a Web site has collected about them, including a reasonable opportunity to review information and to correct inaccuracies or delete information.

- Security. Web sites would be required to take reasonable steps to protect the security of the information they collect from consumers.

If these regulations are to do any good, they must address harms caused by the use of information. In approaching this issue, it is useful to think about the information market as *two markets*: the market for acquisition of

[143] All definitions from FTC's 2000 Statement to Congress.

information and the market for sale and use of information. If any harm results to consumers, it must occur in the second market, where information is used (or conceivably misused). Many of the proposed fair information practices involve only the first market, however. This is particularly the case with notice and choice. Access also has little or nothing to do with the way information is used once it is acquired. Only security deals with information at the point at which harm might occur.

NOTICE

Notice would seem to be straightforward and relatively innocuous. Indeed, most Web sites already offer notice in some form. The FTC survey found that 77 percent of Web sites (weighted by viewers) do so.[144] This might suggest that the costs of a mandatory notice requirement would be minimal. This is not the case, however.[145]

The only way in which requiring notice can affect misuse of information by a particular Web site is by excluding the information from any use, since notice has nothing to do with the way in which information is used once it is acquired. The consumer only has the choice of whether or not to patronize the site. Thus, evaluating the benefits of a mandatory notice requirement involves evaluating the benefits and costs of the use of information *per se*. As discussed previously, there are substantial benefits from the use of information, and little or no evidence of harm.

Moreover, experience with existing mandatory notice requirements is not encouraging. Consumers are now receiving notices from financial institutions required by the recently enacted Gramm-Leach-Bliley ("GLB") financial institution law and are finding them confusing and, in general, not useful.[146]

The effects of mandatory notice requirements are, however, likely to be more serious than the creation of a lot of virtual paperwork. Such requirements can adversely impact innovation in the development both of new uses of information and new products.

[144] Federal Trade Commission, *Privacy Online: Fair Information Practices in the Electronic Marketplace*, May, 2000, available online at http://www.ftc.gov/os/2000/05/index.htm#22 at 15.

[145] This discussion does not apply to a voluntary notice regime, such as exists now, which has the potential to provide much greater flexibility.

[146] John Schwartz, "Privacy Policy Notices Are Called Too Common and Too Confusing," *The New York Times*, May 7, 2001.

Loss of New Uses of Information

Recall that one of the characteristics of information is that, once produced, it can be used multiple times at a very low marginal cost. A mandatory notice requirement would make it difficult, if not impossible, to take advantage of new beneficial uses of information that may become available.

This is because a notice requirement is likely to require that consumers be notified whenever information is used for a purpose not initially intended. This is the case with the financial privacy provisions of the GLB law. GLB establishes the general scope of a consumer notice, but grants the banking regulatory agencies and the FTC authority to issue rules further defining their content. Under GLB, all changes of information use by a financial institution require a new notice.

Similarly, questions have been raised concerning whether on-line merchants who go bankrupt can sell their customer lists for uses not initially intended.[147] Current rulings are ambiguous on this, even though there is no obstacle to brick-and-mortar merchants selling such lists.[148] The difficulty is in the interpretation of the notice requirement.

The FTC report[149] indicates that Web sites should always inform consumers of material changes in their information practices, and that "in some instances, affirmative choice by the consumer may be required." Even without a consent requirement, a notification requirement would be onerous. Consumers who initially supplied information would be difficult to contact and may no longer patronize the particular Web site. In the end, the Web site would be faced with the task of identifying and culling out those individuals who could not be notified. This would diminish the value of the database and probably make the whole process too costly to be worthwhile. Obviously, a consent requirement would make it even more difficult.

There are many examples of ways in which information is now being used that were not contemplated when the information was collected, and which might be illegal if a mandatory notice requirement had been in place. Two examples are digital verification and market segmentation.

[147] Susan Stellin, "Dot-Com Liquidations Put Consumer Data in Limbo," *New York Times*, December 4, 2000.
[148] TRUSTe is curently seeking comment on policies with respect to "Personally Identifiable Information Used in Mergers, Acquisitions, Bankruptcies, Closures, and Dissolutions of Web Sites." http://www.truste.com/bus/spotlight.html.
[149] At 26 in its 2000 Report.

Example: Digital Verification and Authentication[150]

Today, a growing number of high-value transactions on the Internet are made more secure through systems that verify and authenticate individuals. These processes range from simple address verification systems to complete digital certificates that authenticate electronic signatures. These processes require highly specific knowledge of the consumer, because consumers are expected to recall information only they are likely to know. Their answers to specific questions are then matched against a comprehensive database. Databases used for authentication include credit reporting systems that have been developed over the past 30 years to support risk analysis by lenders and others. Credit grantors contribute most of the information contained in these databases. The information comes from loan applications and also from monthly reports about the credit grantor's own experience with the consumer. Much of the data used in authentication systems, such as mortgage information, is more than ten years old. Verification requires the use of databases containing, among other data elements, highly accurate name and address tables. These databases have been developed to support direct marketing services.

The notice requirements now evolving through legislation and regulatory actions would have led to unintended harmful consequences. Assume that such notice systems were in place ten years ago and governed the collection and use of information in databases now used for digital authentication and verification. The notices would have required specific, detailed statements regarding what information was being collected and exactly how it would be used. This notice would have constituted a "contract" between the company administering the database and the consumer. Violation of this notice would be actionable either by FTC enforcement or a private right of action. Credit grantors could not have anticipated that information provided to consumer reporting agencies would some day be used to authenticate consumers on the Internet – a medium which did not even exist as we know it today. The same goes for marketers, retailers and cataloguers who shared or contributed information to databases that are now used to verify individuals.

Moreover, the possibility of forbidding use of information for verification purposes is not hypothetical. The European Union Directive on the Protection of Personal Data is much more stringent than current U.S.

[150] This analysis is based in part on information provided by Tony Hadley and Marty Abrams of Experian. See also Experian, E-series White Paper Authentication, February 2001, available at http://www.experian.com/eseries/authentication_whitepaper.html.

policies, and makes it much more difficult to use data for purposes other than those for which it was initially gathered. As a result, it is generally impossible under the EU Directive to use address information to verify the identity of a credit card user. Therefore, credit fraud is much more common in Europe, and credit cards are more difficult to obtain and more expensive.[151]

Example: Market Segmentation

Information is used to tailor products and advertisements to consumers who will find them particularly useful. This market segmentation is quite valuable both to consumers and to companies. But often the information used in determining the value of products to consumers is information that has been gathered for other purposes. Had provisions like those now contemplated under mandatory notice systems been in place earlier, then it might now be impossible for firms to use information in this way. The result would be that consumers would be substantially worse off– but not in ways that would be easily measurable or detectable. Consumers would not even be aware of the opportunities lost due to an overly stringent interpretation of the law.

Loss of New Uses of the Internet

Over the long run, regulation is likely to hinder potential new uses of the Internet. We would never know which innovative new uses did not come into being because they were inconsistent with regulation. Consumers did not, for example, miss the new technologies associated with telephony that were stopped or delayed by excessive FCC regulation until they came about as a result of deregulation.

Example: Handheld Units

Handheld units provide an example of a product whose development may be hindered by a notice requirement. New handheld units, including personal digital assistants (PDAs), and Web-enabled cell phones are now able to

[151] Solveig Singleton, "Privacy and Human Rights: Comparing the United States to Europe," in Competitive Enterprise Institute, *The Future of Financial Privacy*, Washington, 2000.

access the Internet. It is also possible to download Web sites to handheld units in the process of synchronizing them with a user's desktop computer. There are even new technologies that may make Web information available through audio means. According to some forecasts, the future of the Web is in numerous small portable specialized Web-enabled devices that will be able to communicate with each other and with the Web.[152]

The interaction of these new technologies with privacy policies, including notice requirements is problematic. Notifying users of privacy policies on a PDA or mobile telephone would be difficult, because the screens are too small and too slow to display meaningful notice information. Having notice policies read aloud by an audio-enabled Web site would also be impractical.

Mobile phones or PDAs with GPS chips will be able, using the Internet, to track the geographic location of individual consumers.[153] This will enable better delivery of emergency services to injured persons. The chips will also enable individuals to obtain personalized information relevant to their location, such as driving directions or the location of restaurants or movies. General Motors is planning to use this technology to send information to users of its OnStar vehicle- based navigation system.[154]

Privacy issues are important with these devices. Palm is developing an opt-in program for location chips. DoubleClick will not begin delivering ads until privacy issues are worked out. TRUSTe is developing standards for privacy policies. Of course, the difficulty of presenting privacy policies on small screens applies to these devices as well.

There are at least two lessons from the story of this new technology. First, industry is already responding to privacy concerns. Second, a government-mandated privacy policy could retard or even stop the technology's development. For example, a law mandating notice and standards for notice could be inconsistent with the size of screen available, and certainly with audible Web sites. Consumers would then lose the benefits of a valuable technology. This potential loss, should it occur, would not even be recognized, because innovations that do not occur are not missed.

Moreover, it is not fanciful to fear that regulations with a minimal impact on one communication medium can have major impacts on another. There are at least two recent examples of this:. advertisements for pharmaceuticals on television, and broadcast ads for credit suppliers.

[152] Kevin Maney, "Web develops amazing new tangles," *USA Today*, March 1, 2001.

[153] Discussed, for example, in Anick Jesdanun, "Wireless Tracking Device Coming Soon," AP, October 29, 2000 and Pui-Wing Tam, "...Know Where We Are," *Wall Street Journal*, November 13, 2000.

[154] Rachel Konrad, "General Motors to 'push' ads to drivers," CNET News.com, January 8, 2001.

Example: Pharmaceutical Ads

Until recently, the Food and Drug Administration required that all ads for pharmaceuticals, including ads aimed at consumers, include a "brief summary of prescribing information." This document is the page of warnings and technical discussion in extremely small type following any ad for a pharmaceutical in a magazine or newspaper. The FDA's requirement that this information also be included in television advertising meant that such advertising was virtually nonexistent.[155] The recent change in FDA policy with respect to such advertising has caused an increase in pharmaceutical advertising on TV. Until the change, however, consumers were denied the benefits of these ads, because a policy designed for print media was carried over to broadcast where the policy did not work.[156]

Example: Truth-in-Lending

Similarly, the Truth in Lending Act[157] includes the following "requisite disclosures in advertisement:"

> If any advertisement to which this section applies states the amount of the downpayment, if any, the amount of any installment payment, the dollar amount of any finance charge, or the number of installments or the period of repayment, then the advertisement shall state all of the following items:
>
> 1. The downpayment, if any.
> 2. The terms of repayment.

[155] It was legal to either name a drug or list a condition for which a drug would be appropriate without including the summary, but it was not legal to provide information about both in an ad without including all of the mandated disclosure information.

[156] These issues are discussed in Paul H. Rubin., "Ignorance is Death: The FDA's Advertising Restrictions," in Roger D. Feldman, Editor, *American Health Care: Government, Market Processes, and the Public Interest*, The Independent Institute and Transaction Publishers, 2000, 285-311.

[157] Truth in Lending Act (15 U.S.C. §§ 1601-1667f, as amended), Sec. 1664 (d), Advertising of credit other than open end plans.

3. The rate of the finance charge expressed as an annual percentage rate.

Thus, the act requires that any mention of the down payment, the amount of payment, or other variables "triggers" disclosures of the other terms. This means that radio ads cannot give any terms of repayment without providing virtually all of them. If a radio ad says "nothing down, 36 months to pay," it must also give terms of repayment and the finance charge. The result is that radio ads for many credit-granting institutions (and particularly for automobile dealers) use generalities such as "easy credit terms" or "low down payment." The required disclosures are straightforward in print media, but become onerous in broadcast media.

In both of these cases, regulations were written to govern information disclosure in one medium that appeared reasonable for that medium. When extended to another medium, however, the regulations became onerous and had the effect of denying valuable information to consumers. Increased regulation of Internet privacy could have a similar effect– making it difficult or impossible for Web sites to be used on handheld devices. Moreover, since the Internet is new and rapidly changing, such regulations could easily have unintended consequences for uses that we do not now know about.

Voluntary Disclosure

Consumers do not need a mandatory notice requirement to have information about Web sites' privacy policies. Competitive forces will assure that information is provided and many Web sites are now voluntarily doing so. If consumers care about privacy, it pays some Web sites to advertise the benefits they provide and also to inform consumers of less-than-satisfactory practices of their competitors.

Privacy is a dimension of product quality, and (as Grossman[158] has shown) all Web sites except those providing the lowest level of privacy will have an incentive to disclose their privacy policy. Even if Web sites would prefer not to disclose, competitive forces will require them to do so. Noam also makes this point with respect to personal information.[159] In fact, we now see that

[158] Sanford Grossman , "The Informational Role of Warranties and Private Disclosure about Product Quality," *Journal of Law and Economics* v. 24, December 1981, pp. 461-483.

[159] Eli Noam, "Privacy and Self-Regulation: Markets for Electronic Privacy," in Privacy and Self-Regulation in the Information Age, U. S. Department of Commerce, Washington, 1997, http://www.ntia.doc.gov/reports/privacy/selfreg1.htm.

some Web-based businesses are beginning to advertise that they do provide more privacy than others.[160]

CHOICE

Choice also involves the market for acquisition of information, not the market for the sale and use of information, which is where potential harms may occur. As with notice, the only way choice can reduce misuse of information is by preventing information from being used at all. Recent estimates suggest the cost of this is high.

Michael A. Turner[161] estimates that opt-in requirements would reduce the data available to catalog apparel retailers at a cost of one billion dollars (in a $15 billion dollar market). Restricting the use of gathered data from non-customers would increase costs by 3.5 to 11 percent.

An Ernst and Young study conducted for the Financial Services Roundtable[162] estimates that current information sharing by the 90 largest financial institutions provides benefits of $195 per household per year, for a total saving of $17 billion. Of this, $9 billion are from information sharing with third parties and $8 billion from information sharing with affiliates.[163]

There are two issues with respect to choice. The first concerns the choices that are made available to consumers. The second concerns the method of choice.

The various FTC documents are ambiguous as to what choices should be made available to a consumer. In particular, it is unclear whether the FTC wants to require that services be made available on the same basis regardless of whether a consumer chooses to share information. For example, the FTC states:[164] "Similarly, the agency could examine the specific contours of the Choice requirement, particularly its application to programs in which the sole reason for providing consumers a particular benefit is the collection and use of personal information (e.g., providing discounts to consumers expressly

[160] Alec Klein and Shannon Henry, "On Reflection, a Puzzling Ad Campaign," *The Washington Post*, March 1, 2001, p. E1.

[161] "The Impact of Data Restrictions on Consumer Distance Shopping, available at http://www.the-dma.org/isec/9.pdf.

[162] "Customer Benefits from Current Information Sharing by Financial Services Companies"

[163] Recall also the estimate, discussed in Chapter 2, that accurate credit reporting reduces interest costs by 200 basis points per year on average, which translates into $4,000 on a $200,000 mortgage, or $85 billion to $100 billion annually. Credit reporting would not be possible if credit information were not available.

[164] At 37 in its 2000 Report.

conditioned on the exchange of personal information.)" Note that the FTC is expressly ambiguous on how this transaction would be treated; the agency might decide that all consumers should be entitled to a particular choice, even those who withhold their personal information.

This would likely make not viable a class of business models that depend on revenues from advertising and the sale of information. Even consumers who would willingly exchange their information would not have the choice available. Particular goods and services would become unavailable and consumers would, in a very real sense, be deprived of choices. Discount cards at grocery stores, free Web browsing, free Web "wallets,"[165] free e-mail, free customizable homepages, and many other goods that are now financed through sale of information could easily be forced out of the market. The services might continue if a sufficient number of consumers chose to continue providing information, but those who chose not to provide information would free ride on others.

METHOD OF CHOICE: OPT-IN OR OPT-OUT

These terms deal with the default for use of information. Under an "opt-in" rule, an individual must affirmatively agree to have his information used. Under "opt-out," the gatherer of the information has the right to use it unless the individual requests that it not be used.

Despite the current debate concerning which rule is better, there is no need for regulators to make this decision *a priori*. In some contexts, Web site managers may choose opt-in because of the type of information at issue, the desires of the Web site's customer base or the value of the information to the Web site operator. Recall from Chapter 2 that some advertisers are willing to pay more for opt-in lists than for opt-out lists, on the theory that opt-in lists contain consumers who are more interested in receiving advertising messages. Because different practices will be best for different situations, Web site managers should be able to configure their sites as they want, and consumers decide which sites to visit.

Moreover, it would be very difficult to design a regulatory regime that takes into account the complexities of the information flows in this market (see Figure 1 in Chapter 2). Theoretically, Web sites could be required to offer consumers choice with respect to how every single piece of information

[165] Web based agents that automatically fill out ordering or registration forms with consumer information, such as Dash.com.

is used. This would clearly involve a lot of choices and would place a significant burden on activities on the Internet.

The FTC report makes little mention of this issue. (The Commission does indicate that that it is in favor of opt-in for medical records.[166]) The NAAG proposal is more explicit. It advocates a policy of opt-in for the use of "personally identifiable information for purposes other than the purpose for which the information was obtained." For anonymous or aggregate data, NAAG would accept opt-out.

The available evidence indicates that the great majority of consumers accept the default. In testimony before the FTC on the experience of one firm, a witness indicated that, when the default was opt-in, 85 percent of consumers chose not to provide their data. In contrast, 95 percent chose to provide their data when the default was opt-out.[167] Thus, requiring opt-in would dramatically reduce the amount of information available to the economy and would impose substantial costs on consumers.

Why do consumers accept the default? It may be because they don't consider the issue very important one way or the other. But, it is more likely that the transactions costs associated with making a decision, which include reading a detailed notice and understanding the nature of the choice, are not trivial. That is, the costs associated with the process of opting-in or opting-out are not insignificant.

The following example (from a proponent of opt-in) indicates the sort of transactions costs associated with opt-in:[168]

> Evidence on how companies behave in an opt-in environment suggests that such a system may be more efficient for consumers than the current system. After the FCC ruled that phone companies seeking to use phone-calling patterns for marketing purposes must first obtain the consumer's permission, the telephone company in my area attempted to secure that permission. Its representatives called and sent mailings to subscribers. The company also set up a toll-free number for consumers with questions. The mailing I received was brief, printed in different colors, and

[166] Note 21 in its 2000 Report.

[167] Testimony by Parry Ponemon, PriceWaterhouseCoopers, at the FTC hearing, "Wireless Web, Data Services and Beyond: Emerging Technologies and Consumer Issues," December 12, 2000, Vol. 2, p. 232.

[168] Jeff Sovern, "Opting In, Opting Out, Or No Options At All: The Fight for Control of Personal Information," 74 *Washington Law Review* 1033, October 1999, at 1102, footnote omitted.

> written in plain English. It also promised, in words which were underlined, that "we'll never share this information with any outside company." A postage-paid envelope and a printed form were included for consumers to respond. Consumers who accept the offer need only check a box, sign and date the form, and print their name. The company also offered consumers incentives to sign up – such as a five-dollar check, two free movie tickets, or a ten-dollar certificate from certain retailers – thus increasing the likelihood that consumers will pay attention to the information. In sum, the company has done everything it can to eliminate consumer transaction costs.

While this procedure may have reduced "consumer" transactions costs, it clearly increased total transactions costs.

Similarly, a US West telemarketing campaign was only able to obtain an opt-in rate of 29 percent among residential subscribers, and at a cost of $20.66 per positive response.[169] These transactions costs are ultimately paid by consumers. Consumers also incur direct nuisance costs being at the receiving end of such campaigns.[170] These types of costs are likely to multiply rapidly as there may be numerous requirements for opt-in consent within a specific category and between categories.

If transactions costs are sufficiently low, then legal rights move to their highest valued uses.[171] The original allocation of the right does not matter. If transactions costs are significant, however, then it is efficient to give the right to the party who values it the most, or the party who would buy it if transactions costs didn't get in the way.[172] In this case, that party is the Web site. This implies that, if there is to be any requirement at all, it should be opt-out.

This is because the purpose of obtaining information about consumers is to provide them with targeted advertising – advertising of products likely to be of use to them. This is exactly the transaction that consumers indicate

[169] Cited in Michael A. Turner, "The Impact of Data Restrictions on Consumer Distance Shopping, available at http://www.the-dma.org/isec/9.pdf.

[170] Discussed in Fred H. Cate and Michael E. Staten, "Protecting Privacy in the New Millennium: The Fallacy of 'Opt-In,'" Information Services Executive Council, available at http://www.the-dma.org/isec/optin/shtml.

[171] Ronald H Coase, "The Problem of Social Cost," 3 *Journal of Law and Economics* 1, 1960. Professor Coase was awarded the Nobel Prize in Economic Science in 1991 for this work.

[172] E.g. Richard A. Posner, *Economic Analysis of Law*, Aspen Law and Business, 5th Edition, 1998.

they want to engage in, and so they should be allowed to do so. If transactions costs were low, Web sites would end up with the information. Since transactions costs are not low, efficiency argues for giving the initial right to businesses – that is, for opt-out. Noam[173] and Varian[174] also conclude that opt-out is the most efficient pattern. If the default were opt-in, information would not flow to its highest valued uses. This loss in information would be quite costly, and would lead to price increases as firms would charge higher prices to compensate for the loss of information.

Moreover, it is not correct to argue that opt-out is equivalent to giving a property right to the Web site and opt-in gives it to the consumer. The consumer unambiguously has the initial property right, in that she can choose to opt-out or refuse to visit sites that use opt-out. Sites that use opt-out will charge lower prices, but the right rests with the consumer.

ACCESS

Access allows a consumer to observe his information and perhaps correct errors. This is useful for information that is used personally, such as information used for credit checks and insurance applications. Its usefulness for information used for marketing purposes is much less clear.

The Commission itself has thus far been unable to define access in a meaningful way, despite its inclusion as one of the four basic fair information practices. In its recommendation, the Commission referred to the report of the "FTC Advisory Committee on Online Access and Security."[175] But, as the FTC itself indicated, this committee was unable to reach any substantive conclusions: "The Advisory Committee Report acknowledges that implementing the fair information practice principle of access is a complex task, and there was considerable disagreement among members as to how

[173] Eli Noam, "Privacy and Self-Regulation: Markets for Electronic Privacy," in Privacy and Self-Regulation in the Information Age, U. S. Department of Commerce, Washington, 1997, http://www.ntia.doc.gov/reports/privacy/selfreg1.htm.

[174] Hal Varian, "Economic Aspects of Personal Privacy," in Privacy and Self-Regulation in the Information Age, U. S. Department of Commerce, Washington, DC., 1997 http://www.ntia.doc.gov/reports/privacy/selfreg1.htm.

[175] Federal Trade Commission (2000a), *Final Report of the FTC Advisory Committee on Online Access and Security*, May 15, 2000, http://www.ftc.gov/acoas/papers/finalreport.htm. A fair reading of the report is that the Committee was strongly opposed to regulation of access and security. We indicate some of the areas in which it was opposed, but reading the Report leads one to wonder why the FTC commissioned it if the plan was to proceed independently of the Committee recommendations, which is exactly what occurred.

'reasonable access' should be defined, including whether access should vary with the use or type of data."[176] Indeed, the Advisory Committee Report reads like a guide to unexplored wilderness. It begins on page 3 by indicating that even the definition of access is uncertain: "does it mean the ability to review the data or does it include authority to challenge, modify, disable use, or delete information?" It then asks three additional fundamental questions: "Access to what?" "Who provides access?" and "How easy should access be?" The rest of the report clarifies the nature of these and related issues regarding access, but reaches no conclusions. Thus, the FTC is recommending legislation giving it authority to regulate access, even though the agency acknowledges that it does not know how this should be done.

Consumers may value access to personal data because these data are sometimes used to make decisions about them. For example, data held by credit agencies are used to make decisions regarding granting of credit, and so it is useful for consumers to have access to these data in order to correct them, as is now possible under the Fair Credit Reporting Act.[177] But most of the information held by commercial Web sites is not used in this manner at all. Rather, most of the information is anonymous, used for marketing purposes. The value of accessing these data is small, because the cost of erroneous information is trivial. If a Web site has an individual's name and e-mail address on a list of those considering purchase of a computer, for example, and that individual is not now considering such a purchase, then she may get superfluous e-mail messages or irrelevant banner ads trying to sell a computer. There is a cost to deleting or ignoring these messages– but this cost, in terms of time and nuisance, is less than the cost, in terms of time and nuisance, of going to the Web sites that have the erroneous information and correcting it. Thus, for the sorts of commercial Web sites considered here, consumer access to data is of very limited value.

However access is defined, requiring Web sites to develop mechanisms for consumers to access their personal data would be extremely costly and would likely be unworkable. A recent study by Robert Hahn of the costs of online privacy regulation that specifically focuses on access finds that costs "easily could be in the billions, if not tens of billions of dollars."[178]

[176] From the FTC Report, p. 29, footnote omitted.

[177] Available Online from the FTC website at http://www.ftc.gov/os/statutes/fcra.pdf.

[178] Robert W. Hahn, "An Assessment of the Costs of Proposed Online Privacy Legislation," May 7, 2001, available at http://www.actonline.org/press_room/releases/050801summary.asp. This study has been criticized by Peter Swire. See May 9 press statement, http://www.osu.edu/units/law/swire1/pshome1.htm.

Moreover, as we discuss in the following section, there is a fundamental tension between access and security. Providing greater access reduces security and makes crimes like identity theft more likely.

SECURITY

Security does deal with misuse of information at the point where harm might occur. Thus, cost-effective security requirements might well be justified. There are, however, several reasons for doubting that regulation in this area would be cost-effective. First, there is no reason to believe that businesses now are spending too little on security, since they bear most of the cost of violations. Second, the FTC discussion of security does not indicate that the agency knows how to improve security. Third, the government in general is unable to guarantee the security of its own data, as indicated in numerous GAO studies:

> Computer-supported federal operations are also at risk. Our previous reports and those of agency inspectors general, describe persistent computer security weaknesses that place a variety of critical federal operations, including those at IRS, at risk of disruption, fraud, and inappropriate disclosure. This body of audit evidence led us, in 1997 and again in 1999 reports to the Congress to designate computer security as a government-wide high-risk area. It remains so today.[179]

The FTC indicates that there was more agreement among the members of the Advisory Committee on matters of security than on access, but that there is still a good deal of uncertainty about the issue. Indeed, the FTC argues that if given the authority to enforce security requirements, it would define an "appropriateness" standard through "case-by-case adjudication." In other words, the FTC would not even attempt to define a standard *ex ante*, but rather would sue firms that suffered security violations, and would use this process to define a standard. This case-by-case adjudication would be administered by FTC staff with limited knowledge and understanding of the appropriate technologies. About as specific as the Advisory Committee got was to recommend that "[t]he security program should be appropriate to the

[179] Government Accounting Office, *Information Security: IRS Electronic Filing Systems*, February, 2001, p. 4, footnotes omitted.

circumstances."[180] This standard obviously gives little guidance to firms. Moreover, it seems odd to expect firms to adopt what the FTC views as appropriate levels of security if the FTC cannot specify the parameters of this level in advance. Additionally, as the Advisory Committee indicates, "...both computer systems and methods of violating computer security are evolving at a rapid clip, with the result that computer security is more a process than a state."[181] Thus, the FTC would follow a policy of suing firms that it believes have violated an amorphous and continually changing standard. The Advisory Committee indicated that enforcement of security by the government (the FTC) was one option, but was explicitly unwilling to endorse this as the best option.[182]

The FTC recommended regulation in part because it found that there was inadequate discussion of security issues in the privacy notices associated with Web sites. But the Advisory Committee states that:

> Since it is difficult to convey any useful information in a short statement dealing with a subject as complex as the nuts and bolts of security, most such notices would be confusing and convey little to the average consumer. Further, providing too many technical details about security in a security notice could serve as an invitation to hackers. As was discussed at some length by the Committee, these considerations also mean that it is not possible to judge the adequacy of security on Web sites by performing a 'sweep' that focuses on the presence or absence of notices.[183]

It was exactly such a sweep that was used by the FTC to recommend increased regulation.

The Advisory Committee also pointed out that there is a fundamental conflict between access and security. Increasing ease of access makes misuse of data for illegal purposes easier, and thus compromises security: "...privacy is lost if a security failure results in access being granted to the wrong person – an investigator making a pretext call, a con man engaged in identity theft, or, in some instances, one family member in conflict with

[180] Advisory Committee Report, at 33.
[181] At 24.
[182] At 34. The other options discussed are: "Rely on Existing Enforcement Options"; "Third-Party Audit of Other Assurance Requirements;" and "Create Express Private Cause of Action." FTC Advisory Commission Report, at 34.
[183] At 27.

another."[184] If a thief had enough information about some individual to fool a Web site which contained additional information about that individual into granting the thief access, then he might be able to engage in identity theft. Therefore, online merchants would be in an impossible position if they tried to provide both access and security. If access were easy, security could be compromised, but if access were difficult, it could be claimed that firms were not providing sufficient access. Presumably the FTC would accept some tradeoff between them, but the details of the tradeoff are unknown. Until they became clear (if they did), every firm would be subject to an enforcement action for providing either too little security or too little access.

Firms already have incentives to provide security, as the FTC Advisory Commission recognized in Section 3.1:

> Security is contextual: to achieve appropriate security, security professionals typically vary the level of protection based on the value of the information on the system, the cost of particular security measures and the cost of a security failure in terms of both liability and public confidence.

If security were mandated and enforced by the FTC, then the FTC would be in the business of second guessing these decisions. While the FTC in performing its consumer protection mission is better than many other regulators, it nonetheless has incentives to overregulate in this area.[185] For example, if there should be a privacy violation leading to a breach of security, the FTC would tend to underweigh the gains to the firm from the behavior that caused the violation and overweigh the losses, and so penalize a firm for what might have been *ex ante* efficient behavior.

EFFECTS ON COMPETITION

Any regulation that raises the costs of advertising and obtaining customer lists would have an adverse effect on new entrants. Regulation would reduce advertising, and advertising typically benefits new entrants and small firms more than large, established firms.[186] This is particularly true for Internet

[184] At 19.
[185] Paul H. Rubin "Economics and the Regulation of Deception," *Cato Journal*, 1991,667-690, 1991.
[186] John E. Calfee, *Fear of Persuasion: A New Perspective on Advertising and Regulation*, American Enterprise Institute, Washington, 1997.

advertising, where established firms have lists of their own customers and visitors to their Web sites, but new firms must purchase such lists. As long as there is a market for customer lists and other such information, entrants can begin competing relatively easily. However, if regulation should reduce the size of this market and increase costs, competition from new entrants would be reduced.

Even notice requirements would disproportionately impact small businesses, because the costs of compliance are largely fixed. Firms operating Web sites would likely find it necessary to hire an attorney either to write a notice about privacy policies or to otherwise assure that the firm was in compliance with the requirements. These firms are already finding it useful to hire privacy officers.[187] Allowing access and enforcing security regulations would add to the costs. Most of these costs are "fixed" costs, and so are higher per unit of output for small as compared to large firms. Thus, any such regulations would serve at least in part as a barrier to entry against small firms, and as a source of protection for large established firms.[188]

Small startup companies with new ideas and new business models have been particularly important in the Internet economy. Such companies are often started by one or two "techies" – people with an interest in technology. These people probably would be unlikely to have a lawyer on retainer. Privacy regulations are likely to be particularly harmful in this environment.[189]

FEDERAL PREEMPTION

The overall conclusion of our study is that regulation is not called for and would be costly for the economy. Given the nature of the Internet, regulation at the state level has the potential to produce additional costs, because it would interfere with interstate commerce.

The issue of federal preemption of state regulatory authority is part of many legislative proposals. Indeed, many businesses might be willing to

[187] Kemba J. Dunham, "The Jungle: Focus on Recruitment, Pay and Getting Ahead: A New Playing Field," *The Wall Street Journal* March 20, 2001.

[188] For the general point that firms may gain when costs of competitors increase, see Steven C. Salop and David T. Scheffman, "Raising Rivals' Costs," 73 *American Economic Review* (Papers and Proceedings) 267 1983.

[189] Discussed in Peter P. Swire and Robert E. Litan, *None of Your Business: World Data Flows, Electronic Commerce, and the European Privacy Directive*, Washington: Brookings Institution Press, 1998, at 78-79.

accept some federal regulation of privacy if they can avoid fifty separate state laws dealing with the issue.

Not surprisingly, the NAAG advocates that federal law should not preempt state regulation. Adopting the NAAG position would be troublesome. States do have independent authority to enforce consumer protection laws, and the NAAG argues that this authority should carry over to laws regulating the Internet. However, given the nature of the Internet and of commerce on the Internet, it is difficult to see how such laws would work without impeding interstate commerce.[190] Moreover, the states, in enforcing consumer protection laws, are less likely than the Federal Government to use economic analysis to inform their legal strategies, and therefore often end up causing consumer harm.[191]

It might appear that competition among states would limit the costs of regulation. However, experience with other state regulations suggests that states tend to adopt regulations costly to producers in other states in order to provide benefits to consumers in the regulating state. Similar problems occur when states impose excessive or inefficient product liability rules; citizens in the rest of the country are implicitly forced to subsidize consumers in the more regulatory state.[192]

SUMMARY

The potential benefits of privacy regulation are extremely limited, because there is no evidence of consumer harm. On the other hand, privacy regulation would impose substantial costs on the operation of the information economy.

Privacy regulations would be subject to the same problems associated with many other regulatory programs. They would be inflexible and difficult to change. This would be a particular problem as applied to the Internet, which is constantly changing. Moreover, consumers have very different

[190] For a contrary argument in favor of state regulation, see Bruce H. Kobayashi and Larry E. Ribstein, "A State Recipe for Cookies: State Regulation of Consumer Marketing Information," American Enterprise Institute, Washington, January 16, 2001, available online at http://www.federalismproject.org/conlaw/ecommerce/cookies.html. For their proposal to work, courts would have to enforce contractual choice of law and forum; this would require a major change in the behavior of courts with respect to contracts accepted by consumers at the time of purchase.

[191] Discussed in Rubin, 1991.

[192] Paul H. Rubin, John Calfee and Mark Grady, "*BMW vs Gore*: Mitigating The Punitive Economics of Punitive Damages," *Supreme Court Economic Review*, 1997, 179-216.

preferences concerning privacy. Costs would be imposed on those consumers whose preferences did not match the new regulatory requirements.

Privacy regulations would "crowd out" private market responses of the sort that are developing in response to privacy concerns. They would also be anticompetitive and disproportionately costly to small businesses and new entrants.

Virtually all proposals for privacy regulation incorporate some variant of the FTC's four "fair information practices:" Notice, Choice, Access and Security. Each of these requirements would impose significant costs on consumers.

While notice sounds relatively innocuous, a mandatory notice requirement would make it difficult, if not impossible, to take advantage of beneficial new uses of information. This would be especially costly, because information, once produced, can be used multiple times at a very low marginal cost.

Notice would also likely interfere with the development of potential new uses of the Internet. For example, industry is responding to privacy concerns associated with new handheld devices, but would have a much more difficult time doing so if faced with a specific government notice mandate.

A choice requirement would ironically deprive consumers of choices. A whole class of business models that provide goods and services to consumers using revenues from advertising and the sale of information may cease to be viable. "Free" goods and services would become unavailable and consumers would, in a very real sense, be deprived of choices.

Regulators should avoid making an *a priori* decision between opt-in and opt-out. The relative merits of each will vary, depending on the circumstances. Moreover, it would be very difficult to design a choice regime that takes into account the complexities of the information flows in this market.

If there is to be any requirement, however, it should be opt-out. Our analysis indicates that an opt-in requirement would dramatically reduce the amount of information available to the economy and impose substantial costs on consumers. Because the transactions costs associated with making an opt-in or opt-out decision are not trivial, the efficient solution is to give the initial rights to the information to the Web site.

Implementing an access requirement would be extremely costly and would reduce the security of data on the Internet. The FTC itself has been unable to define access in a meaningful way. Similar difficulties apply to a security requirement, which the FTC indicates it would adjudicate on a case-by-case basis.

Finally, with respect to the issue of preemption of state regulation, we conclude that, given the nature of the Internet, a variety of different state privacy regulations would not be desirable because they would impede interstate commerce. Moreover, as in some other areas, states would have an incentive to adopt regulations that imposed costs on out-of-state producers.

Chapter 7

Conclusion

The purpose of this study has been to apply standard economic reasoning to the question of whether to impose new regulation on the commercial market for personal information. We have found that the commercial market for information appears to be working well and responding to consumers' concerns. Despite perceptions to the contrary, there is a striking lack of evidence of consumer harm from privacy violations associated with the commercial use of information for advertising and marketing purposes. We have found no evidence of market failure that would justify the adoption of new regulations in this area. Because there is no evidence of consumer harm, there is no reason to expect regulation to produce any benefits.

The commercial information market provides large benefits for consumers. The availability of personal information enables advertising to be matched to consumers' interests. This provides valuable information to consumers and reduces spam and other unwanted advertising messages. Targeted advertising also pays for a variety of services that are provided to consumers on the Internet without charge.

Consumers' concerns about privacy appear to some extent to be due to a lack of understanding about how the advertising information market works. Advertisers are generally not interested in individuals, but rather in large blocks of consumers who share attributes that make them good marketing prospects. Advertising information is manipulated by computers that search for these attributes. Thus, the consumer remains anonymous and metaphors about Web sites tracking particular individuals are misleading.

None of the major proponents of increased regulation has made a convincing case. The FTC, which issued a major report recommending new regulation of information on the Internet, has provided no evidence of actual harm to consumers or of market failure. Regulation imposed in the absence of market failure is likely to do more harm than good.

This study examined the major categories of market failure in some detail and did not find them applicable to the commercial information market. The evidence indicates that consumers are sufficiently informed about privacy policies of Web sites to make informed decisions. Consumers are quick to respond adversely when Web sites undertake activities that violate consumers' expectations. Voluntary "seal" programs are one important means by which consumers obtain information about Web sites information practices. In addition, new technologies that allow consumers to better control their information flows are becoming available.

We also found that there was no evidence of a negative externality, which would indicate that the flow of information should be restricted. On the contrary, the market for personal information is characterized by significant positive externalities, indicating that, to the extent a market failure exists, it results in less information being produced and used than is socially optimal.

Virtually all proposals to regulate privacy on the Internet incorporate some combination of the FTC's four "fair information practices": Notice, Choice, Access and Security. Each of these requirements would raise the cost and reduce the amount of information available, and thereby impose costs on the economy. They would reduce the quantity and quality of information provided to consumers through targeted advertising and the free services that are supported by advertising revenues.

In addition, regulation of the advertising information market would have potentially serious adverse consequences for other sectors. The information that is used by the advertising sector is also used by other important sectors of the economy, such as credit and insurance. Reducing the personal information available for advertising purposes would simultaneously reduce the information available to those sectors. This would increase the costs of providing credit and insurance and reduce economic performance of those sectors.

Even seemingly innocuous requirements, such as notice and choice, are likely to be costly. While many firms are voluntarily providing consumers information about their privacy policies, a mandatory notice requirement would make it difficult to use information in new ways not foreseen when the notice was designed. It would also impede potential new uses of the Internet that are not consistent with the notice requirements. We have seen precisely this type of effect from regulation in other areas.

Our analysis indicates that an opt-in requirement would dramatically reduce the amount of information available. The market should be allowed to evolve to provide different mechanisms where appropriate. If any requirement is imposed, it should be opt-out, which would clearly be less costly.

The information-technology revolution is in a very real sense about reducing the cost of collecting, processing and using information – and in the process making information for many purposes more available. This is clearly a good thing. Information is a valuable commodity that is used in many different ways by many different sectors of the economy.

Regulation imposed on a medium like the Internet that is changing so rapidly would have unpredictable consequences. The costs would take many forms. Regulation could create market failures where none now exist. Perhaps the most serious cost would be a loss of innovation – new uses of information and of the Internet itself that would be frustrated by a new regulatory regime. All this would slow the progress of the IT revolution with potentially adverse implications for growth and productivity.

Glossary

Access: (FTC Definition, 2000 Statement to Congress). Web sites would be required to offer consumers reasonable access to the information a Web site has collected about them, including a reasonable opportunity to review information and to correct inaccuracies or delete information.

Adverse Selection: A situation in which markets have difficulty functioning because only consumers with specific characteristics enter the market. This is a common problem in insurance markets, where only those most likely to need insurance – i.e., those at highest risk – will purchase it. Insurance companies are careful to avoid this problem. Similarly, lenders do not want to attract bad credit risks. A lemons market (see below) is caused by adverse selection. It is an example of a problem associated with asymmetric information (see below).

Asymmetric Information: Information known to one party to a transaction, but not to another. The existence of asymmetric information can cause a "market failure" because parties who know that they have less information may not be willing to engage in trade or exchange. The resulting market is sometimes called a "lemons market."

Banner Ad: The ads that appear at the top of many Web sites. Often these ads are selected or "targeted" based on the particular Web site visited or the past surfing history or other characteristics of the person visiting the Web site. These ads are generally placed by web based advertising agencies, of which DoubleClick is the largest.

Browser and Server: A browser is the program used to surf or browse the Internet. The two main programs are Netscape Navigator and Microsoft Explorer. These programs reside on the computer of the individual surfing the net. The Web sites that the surfer visits reside on a server, a computer that is visited in the browsing process.

Choice: (FTC Definition, 2000 Statement to Congress). Web sites would be required to offer consumers choices as to how their personal identifying information is used beyond the use for which the information was provided (e.g., to consummate a transaction). Such choice would encompass both internal secondary uses (such as marketing back to consumers) and external secondary uses (such as disclosing data to other entities).

Coase Theorem: The fundamental result in law and economics, named for the Nobel prize laureate in economics Ronald Coase, first stated in an article in 1960.[193] This theorem states that if transactions costs are sufficiently low, then it does not matter which party in a transaction is given a particular legal right. The right will move to that party that values it the most, no matter how it is initially assigned. When transactions costs are high, the efficient solution is to assign the initial right to the party that values it most.

Cookie: A bit of code that is transferred to a home computer from a Web site during surfing. This enables the server to identify the computer next time it logs on. If you subscribe to the *New York Times* online, for example, there is a cookie that indicates that you are registered whenever you log on to the *Times* site, so that you do not have to reenter your name and password. Cookies can also be used to keep track of a surfer's history in contacting various Web sites. A "first party cookie" is put on your computer by the site you are visiting. A "third party cookie" is put on by someone else (with the permission of the host site), often an online advertising agency such as Engage.

European Privacy Directive (The European Union Directive on Data Protection): The basic law governing privacy of data in the European Union. An important point is that "[m]ember states shall provide that personal data must be collected for specified, explicit and legitimate purposes and not further processed in a way incompatible with those purposes." This has

Externality: Some result of an action that is not borne by the decision-maker. If there are external costs, decision-makers will undertake too much of an activity, since they do not bear all costs; pollution is the classic example. If there are external benefits, decision makers will undertake too little of an activity, since they do not get all benefits. Information is often associated with external benefits. Existence of externalities can lead to market failure.

[193]Ronald H. Coase, "The Problem of Social Cost," 3 *Journal of Law and Economics* 1, 1960.

Fair Information Practices: Rules governing uses of information. The exact set of Fair Information Practices varies. For example the FTC includes Notice, Choice, Access, and Security; the NAAG adds Enforcement and No Preemption of States.

Federal Trade Commission (FTC): An independent regulatory agency of the Federal Government. The FTC has Bureaus regulating Competition (antitrust) and Consumer Protection. There is also a Bureau of Economics, which provides economic analysis of policies recommended by the other Bureaus. The Bureau of Consumer Protection has studied issues related to privacy, and has made various recommendations to Congress.[194]

FTC Advisory Committee on Online Access and Security: A Committee appointed by the FTC to make recommendations. Its recommendations were in Federal Trade Commission (2000a), Final Report of the FTC Advisory Committee on Online Access and Security, May 15, 2000, http://www.ftc.gov/acoas/papers/finalreport.htm.

Lemons Market: A market failure associated with asymmetric information, where only low quality sellers exist.

Market Failure: Some condition causing markets to deliver suboptimal performance. The classic causes are market power, externalities, and asymmetric information.

The National Association of Attorneys General (NAAG): The association of attorneys general from each state. In proposing and enforcing regulations, this organization sometimes works with the federal government (e.g., the FTC) and sometimes independently.

[194] Federal Trade Commission, *Privacy Online: A Report to Congress*, June, 1998, available online at http://www.ftc.gov/reports/privacy3/index.htm; Federal Trade Commission, *Privacy Online: Fair Information Practices in the Electronic Marketplace*, May, 2000, available online at http://www.ftc.gov/os/2000/05/index.htm#22.

Notice: (FTC Definition, 2000 Statement to Congress). Web sites would be required to provide consumers clear and conspicuous notice of their information practices, including what information they collect, how they collect it (e.g., directly or through non-obvious means such as cookies), how they use it, how they provide Choice, Access, and Security to consumers, whether they disclose the information collected to other entities, and whether other entities are collecting information through the site.

Opt-in and Opt-out: These terms deal with the default for use of information. Under an "opt-in" rule, an individual must affirmatively agree to have information used. Under "opt-out" the gatherer of the information has the right to use it unless the individual requests that it not be used. Privacy advocates are generally in favor of opt-in; that is, they believe that information should not be used unless the consumer specifically allows this use. However, evidence indicates that most consumers accept whatever default exists, so opt-in would lead to much reduced use of information in the economy.

P3P: The Platform for Privacy Preferences, a protocol that will enable users to program their browsers with their particular privacy preferences. Web sites will also be P3P enabled, so that the browser and a website will be able to instantly determine if the preferences of the person browsing match the settings of the website. P3P is still in development, but will soon be available for downloading.

PDAs: "Personal Digital Assistants," small handheld computers used for schedules and other purposes. Palm and the Pocket PC are the leading standards.

Security: (FTC Definition, 2000 Statement to Congress). Web sites would be required to take reasonable steps to protect the security of the information they collect from consumers.

Spam: E-mails sent at random to large numbers of addresses for marketing purposes. Consumers generally dislike receiving spam.

Bibliography

Baker, Edwin C., (1978), "Posner's Privacy Mystery and the Failure of Economic Analysis of Law," 12 *Georgia Law Review* 3 (Spring), 475-496.

Belgum, Karl D., (1999), "Who Leads at Half-time? Three Conflicting Visions of Internet Privacy Policy," 6 *Richmond Journal of Law and Technology* 1 (Symposium), available at http://www.richmond.edu/jolt/v6i1/belgum.html.

Bloustein, Edward J., (1978), "Privacy is Dear at Any Price: A Reponse to Professor Posner's Economic Theory," 12 *Georgia Law Review* 3 (Spring), 429-454.

Brandeis, Louis D., and Warren, Samuel D., (1890), "The Right to Privacy," 4 *Harvard Law Review* 5 (December), 193-220.

Carlton, Dennis W., (1980), "The Law and Economics of Rights in Valuable Information: A Comment," 9 *Journal of Legal Studies* 4 (December), 725-726.

Cate, Fred H., (1997), *Privacy in the Information Age*, Washington, D.C.: Brookings Institution Press.

Caudill, Eve M., and Murphy, Patrick E., (2000), "Consumer Online Privacy: Legal and Ethical Issues," 19 *Public Policy and Marketing* 1 (Spring), 7-19.

Cohen, Julie E., (2000), "Examined Lives: Informational Privacy and the Subject as Object," 52 *Stanford Law Review* 5 (May), 1373-1438.

Coleman, James S., (1980), "An Introduction to Privacy in Economics and Politics," 9 *Journal of Legal Studies* 4 (December), 645-648.

Competitive Enterprise Institute, (2000), *The Future of Financial Privacy: Private Choices Versus Political Rules*, Washington, D.C.: Competitive Enterprise Institute.

Culnan, Mary J., (2000), "Protecting Privacy Online: Is Self-Regulation Working?" 19 *Public Policy and Marketing* 1 (Spring), 20-26.

D'Amato, Anthony, (1978), "Comment: Professor Posner's Lecture on Privacy," 12 *Georgia Law Review* 3 (Spring), 497-504.

Electronic Privacy Information Center (EPIC) and Junkbusters, "Pretty Poor Privacy: An Assessment of P3P and Internet Privacy." http://www.epic.org/reports/prettypoorprivacy.html.

Epstein, Richard A., (1980), "A Taste for Privacy? Evolution and the Emergence of a Naturalistic Ethic," 9 *Journal of Legal Studies* 4 (December), 665-682.

Epstein, Richard A., (2000), "Privacy, Publication, and the First Amendment: The Dangers of First Amendment Exceptionalism," 52 *Stanford Law Review* 5 (May), 1003-1048.

Federal Trade Commission (1998), *Privacy Online: A Report to Congress*, (June) available online at http://www.ftc.gov/reports/privacy3/index.htm.

Federal Trade Commission (2000), *Privacy Online: Fair Information Practices in the Electronic Marketplace*, (May), available online at http://www.ftc.gov/os/2000/05/index.htm#22.

Federal Trade Commission (2000), *Final Report of the Federal Trade Commission Advisory Committee on Online Access and Security* (May), available online at http://www.ftc.gov/acoas/papers/finalreport.htm.

Fried, Charles, (1978), "Privacy: Economics and Ethics: A Comment on Posner," 12 *Georgia Law Review* 3 (Spring), 423-428.

General Accounting Office (2000), *Internet Privacy: Agencies' Efforts to Implement OMB's Privacy Policy*, (September), available online at www.gao.gov.

General Accounting Office (2000), *Internet Privacy: Comparison of Federal Agency Practices With FTC's Fair Information Principles*, (September), available online at www.gao.gov.

General Accounting Office (2000), *Information Security: Serious and Widespread Weaknesses Persist at Federal Agencies*, (September), availabe online at www.gao.gov.

Gould, John P., (1980). "Privacy and The Economics of Information," 9 *Journal of Legal Studies* 4 (December), 827-842.

Hahn, Robert W. (2001), "An Assessment of the Costs of Proposed Online Privacy Legislation, available online, http://www.actonline.org/pubs/HahnStudy.pdf.

Hartmann, Charles J., and Renas, Stephen M., (1985), "Anglo-American Privacy Law: An Economic Analysis," 5 *International Review of Law and Economics*, 2 (December), 133-152.

Hirshleifer, Jack, (1980), "Privacy: Its Origin, Function, and Future," 9 *Journal of Legal Studies* 4 (December), 649-664.

Kang, Jerry, (1998), "Information Privacy in Cyberspace Transactions," 50 *Stanford Law Review* 4 (April), 1193-1294.

Kitch, Edmund W., (1980), "The Law and Economics of Rights in Valuable Information," 9 *Journal of Legal Studies* 4 (December), 683-724.

Kobayashi, Bruce H., and Ribstein, Larry E. (forthcoming 2001), "A Recipe for Cookies: State Regulation of Consumer Marketing Information," *American Enterprise Institute*, available online at http://www.gmu.edu/departments/law/faculty/papers/docs/01-04.pdf.

Laudon, Kenneth C., (1996), "Markets and Privacy," 39 *Communications of the Association for Computing Machinery*, 9 (September), 92-104.

Laudon, Kenneth C. (1997), "Extensions to the Theory of Markets and Privacy: Mechanics of Pricing Information," in Privacy and Self-Regulation in the Information Age, U. S. Department of Commerce, Washington, DC. http://www.ntia.doc.gov/ reports/privacy/selfreg1.htm.

Lessig, Lawrence, (1998), "The Architecture of Privacy," available online at http://cyber.law.harvard.edu/works/lessig/architecture_priv.pdf

Litan, Robert E., (1999), "Balancing Costs and Benefits of New Privacy Mandates," Working Paper (April), available online at http://www.aei.brookings.org /publications/working/working_99_03.pdf.

Litman, Jessica, (2000), "Information Privacy/Information Property," 52 *Stanford Law Review* 5 (May), 1283-1314.

Murphy, Richard S., (1996), "Property Rights in Personal Information: An Economic Defense of Privacy," 84 *Georgetown Law Journal* 7 (July), 2381-2417.

Petty, Ross D., (2000), "Marketing Without Consent: Consumer Choice and Costs, Privacy, and Public Policy," 19 *Public Policy and Marketing* 1 (Spring), 42-53.

Posner, Richard A., (1981), "The Economics of Privacy," 71 *American Economic Review* 2 (May), 405-409.

Posner, Richard A., (1978), "The Right of Privacy," 12 *Georgia Law Review* 3 (Spring), 393-422.

Privacy Working Group (1995), "Privacy and the National Information Infrastructure: Principles for Providing and Using Personal Information," available online at http://www.iitf.nist.gov/ipc/ipc-pubs/niiprivprin_final.html.

Reidenberg, Joel R., (1999), "Restoring Americans' Privacy in Electronic Commerce," 14 *Berkeley Technology Law Journal* 2 (Spring), 771-792.

Rosen, Jeffrey, (2000), *The Unwanted Gaze: The Destruction of Privacy in America*, New York: Random House.

Rubin, Paul H., (1978), "Government and Privacy: A Comment on 'The Right of Privacy'," 12 *Georgia Law Review* 3 (Spring), 505-512.

Samuelson, Pamela, (2000), "Privacy as Intellectual Property?" 52 *Stanford Law Review* 5 (May),1125-1174.

Schwartz, Paul M., (2000), "Internet Privacy and the State," 32 *Connecticut Law Review* (Spring), 815-859.

Singleton, Solveig, (2001), "Privacy in a Free Country: In Search of Reasonable Principles," NPCA Policy Report 243, Competitive Enterprise Institute, available online at http://www.ncpa.org/studies/s243/s243.html.

Singleton, Solveig, (1998), "Privacy as Censorship: A Skeptical View of Proposals to Regulate Privacy in the Private Sector," Cato Policy Analysis 295 (January 22), available online at http://www.cato.org/pubs/pas/pa~295.html.

Solove, Daniel J., (2001), "Privacy and Power: Computer Databases and Metaphors for Information Privacy," working paper available online through www.ssrn.com.

Sovern, Jeff, (1999), "Opting In, Opting Out, or No Options At All: The Fight for Control of Personal Information," 74 *Washington Law Review* 4 (October), 1033-1118.

Stiger, George J., (1980), "An Introduction to Privacy in Economics and Politics," 9 *Journal of Legal Studies* 4 (December), 623-644.

Swire, Peter P. and Litan, Robert E., (1998), *None of Your Business: World Data Flows, Electronic Commerce, and the European Privacy Directive*, Washington: Brookings Institution Press.

Swire, Peter P., (1997), "Markets, Self-Regulation, and Government Enforcment in the Protection of Personal Information," in Privacy and Self-Regulation in the Information Age, U. S. Department of Commerce, Washington, DC. http://www.ntia.doc.gov/reports/privacy/selfreg1.htm.

Varian, Hal, (1997), "Economic Aspects of Personal Privacy," in Privacy and Self-Regulation in the Information Age, U. S. Department of Commerce, Washington, DC. http://www.ntia.doc.gov/reports/privacy/selfreg1.htm.

Volokh, Eugene, (2000), "Freedom of Speech and Information Privacy: The Troubling Implications of a Right to Stop People from Talking About You," 52 *Stanford Law Review* 5 (May), 1049-1124.

Walker, Kent, (2000), "Where Everybody Knows Your Name: Balancing Community, Commerce, and Freedom in the Information Age," *Stanford Technology Law Review* (online: http://stlr.stanford.edu/STLR/Articles/00_STLR_2/index.htm).

Zittrain, Jonathan, (2000), "What the Publisher Can Teach the Patient: Intellectual Property and Privacy in an Era of Trusted Privication, 52 *Stanford Law Review* 5 (May), 1201-1250.

About the Authors

THOMAS M. LENARD is Vice President for Research and Senior Fellow at The Progress & Freedom Foundation, where he is responsible for the overall direction of the Foundation's major research projects. Dr. Lenard joined the Foundation in 1995 as Senior Fellow and Director of Regulatory Studies.

Prior to joining the Foundation, Dr. Lenard was Vice President of a Washington, DC-based economics consulting firm. Dr. Lenard has also served in senior government positions at the Office of Management and Budget, the Federal Trade Commission and the Council on Wage and Price Stability.

Dr. Lenard was a member of the economics faculty at the University of California, Davis, and has been a visiting economist at the Brookings Institution. He has published and testified before regulatory agencies and congressional committees on a variety of economic issues, including electricity regulation, competition in the computer industry, regulation of drugs and medical devices, and postal competition. Recent publications include *Deregulating Electricity: The Federal Role*; *The Digital Economy Fact Book*; and *Competition, Innovation and The Microsoft Monopoly: Antitrust in the Digital Marketplace*. Dr. Lenard received his B.A. from the University of Wisconsin and his Ph.D. in Economics from Brown University. He currently serves as President of the National Economists Club in Washington D.C.

PAUL H. RUBIN is Professor of Economics and Law at Emory University in Atlanta and Senior Fellow at The Progress & Freedom Foundation. He also serves as editor-in-chief of *Managerial and Decision Economics*. Dr. Rubin is a former Vice President of the Southern Economics Association, and a Fellow of the Public Choice Society.

Dr. Rubin has been Senior Staff Economist at President Reagan's Council of Economic Advisers, Chief Economist at the U.S. Consumer Product Safety Commission, Director of Advertising Economics at the Federal Trade Commission, and Vice President of Glassman-Oliver Economic Consultants, Inc., a litigation consulting firm in Washington D.C. He has taught economics at the University of Georgia, City University of New York, VPI, and George Washington University Law School.

Dr. Rubin has written or edited many books, and published over one hundred articles and chapters on economics, law, and regulation, in journals including the *American Economic Review, Journal of Political Economy, Quarterly Journal of Economics, Journal of Legal Studies, Journal of Law and Economics*, and the *Yale Journal on Regulation*, and has also contributed to the *Wall Street Journal*.

He has consulted widely on litigation related matters, and has addressed numerous business, professional, policy and academic audiences. Dr. Rubin received his B.A. from the University of Cincinnati in 1963 and his Ph.D. from Purdue University in 1970.

Index